Researching a Real UFO

A Practical Guide to WCUFO
Experimentation for Young Scientists

Researching a Real UFO

UFO

A Practical Guide to WCUFO
Experimentation for Young Scientists

by

Christopher Lock HonFSAI and Francisco Villate

2016

First Edition: November 2016

ISBN-13: 978-1-7771550-2-5

Billy Meier pictures, copyrighted by Billy Edward Albert Meier and FIGU. They were used with permission from Billy Meier for this independent investigation. In some cases, electronic copies of his negatives were provided to the authors.

Distributed by Ingram

Based on previous investigations by the same authors:

Zahi, Rhal. *Analysis of the Wedding Cake UFO.* US Library of Congress Copyright © TXu 1-875-255

Francisco Villate (Rhal Zahi), and Christopher Lock.
They are Here - Compelling Evidence of Extraterrestrial Ships Present on Earth.
US Library of Congress Copyright © TXu 2-187-158

Related video links:

www.youtube.com/watch?v=6WHqBvBZOqg (A real ET spaceship? (WCUFO))
www.youtube.com/watch?v=JoZKwqptZ2Y (Yes... it is a real UFO (WCUFO))
www.youtube.com/watch?v=_pjcbF1oK8Q (New revelations of an UFO - WCUFO)

Publisher information on: www.rhalzahi.com

*Anyone dedicated to curiosity, experimentation,
and evidence can call himself a scientist.*
— Alex Brown

In fond memory of
Billy Meier case investigation pioneers
Wendelle Stevens and Professor James Deardorff.

Table of Contents

List of Figures

Figure	Contents	Page

Figure	Contents	Page

Figure	Contents	Page

Figure	Contents	Page

Figure	Contents	Page

List of Tables

Acknowledgements

We would like to thank Billy Meier for permitting us to independently investigate some of his photos.

We also express our appreciation to Christian Frehner for his continuous support and help throughout the work.

Thanks also go to investigators and skeptics for their valuable comments and challenges to improve understanding of these investigations, which to a considerable degree led to the writing of this book.

Preface

After publishing the initial detailed investigation of the WCUFO *Analysis of the Wedding Cake UFO* in 2013, some readers felt somewhat threatened by the mathematical formulas. They thought the analysis a little too technical and mathematical, and too difficult or complicated to follow, notwithstanding the simple experimentation suggestions. So we decided to write an easier book with practical experiments that anyone can do with just simple math skills.

This book is the result. In it, we introduce and detail eleven experiments which analyze the WCUFO photos taken by contactee "Billy" Eduard Albert Meier, popularly known as Billy Meier. The photos show the WCUFO hovering above Meier's courtyard and elsewhere in the Swiss countryside near his home. The suggested experiments verify the WCUFO evidence and status, revealing two sizes of WCUFO, 3.5 m and 7 m in diameter. The book demonstrates that anybody with a few simple skills and tools can perform specific and fun scientific experiments and investigations of this strange UFO.

Apart from just coming to grips with these enigmatic photos, we hope this book may function as a possible guidebook or handbook for budding scientists or young students. The eleven practical and scientific experiments could constitute part of a science course for high school or college students. It could be a course book, an additional text, extracurricular material, or used in any way deemed useful by the reader, teacher or student. It is also for the curious, and people interested in pursuing a scientific process to arrive at certain facts or truths.

We suggest students, experimenters, and readers use this book as a research scientist, actually or even just mentally, conducting a step by step program of any or all eleven experiments and analyses. Ideally, to successfully perform these experiments, readers should possess, or have available the basic skills explained in the Introduction; but most importantly, to have curiosity and an open, critical, and skeptical mind ready for experimentation and discovery of some facts and truth concerning the WCUFO.

The book is also intended to help those interested in the Billy Meier case better understand the nature of these WCUFO photographs, and to shed some much-needed understanding of the truth regarding at least some of these famous – and most controversial – UFO photographs.

Introduction

TRICK OR REAL?

Is there any evidence that *real* UFOs, mysterious and unexplained flying machines, exist? Most observed *real* UFOs are distant disk-shaped flying objects, sometimes moving in a manner unexplained by any known flying device. Are there photos and videos of UFOs taken close to a photographer that science can study and verify as being real unknown objects – *real* UFOs – and not tricks of somebody playing with little UFO Models? Yes, there are. This book takes a look at some truly exceptional examples, and how you too can verify they are *real* UFOs.

How can a scientist or layperson analyze a UFO that a contactee claims comes from outer space, and how can he or she be certain it is a real unknown object with peculiar characteristics – a real UFO? Is there real evidence of such an object for the scientist to study? Surprisingly, again the answer is: Yes, there is. Detailed photographic evidence of one type of UFO is available. At first glance, it may appear that these photographs of this object show tricks or hoaxes, but upon careful investigation, we find irrefutable evidence proving they are not tricks or hoaxes. Surprisingly, some of these important details have been sitting there undiscovered or unnoticed for decades in these UFO photographs.

In this book and *They are Here* (Zahi and Lock) we investigate and reveal these newly discovered details. Proper investigation conclusively shows that these real, sizeable, claimed extraterrestrial ships do exist; and this book gives you the tools to make *your* tests, investigations, and conclusions on the unique photographic images of this polemical UFO: the WCUFO.

"Billy" Eduard Albert Meier says he photographed this unusual UFO numerous times in his home country of Switzerland, and this book puts to critical test some of these Meier photographs.

This book does not prove these crafts are extraterrestrial or from outer space, though they have some remarkably exotic features and capabilities suggesting an extraterrestrial connection. What this book does show is that the WCUFOs are real large-sized flying objects that have never been satisfactorily explained away as anything other than what Meier says they are: Plejaren flying devices, or actual unidentified flying objects: Real UFOs.

THE MEIER CASE

Standing head and shoulders above any other UFO case, the Billy Meier case is by far the biggest and most important UFO case in recorded human history. It is, therefore, the case most in need of investigation. It is also the case providing the most voluminous and highest quality photographic evidence, including the WCUFO photographs that open a window of opportunity for humanity to observe the real probability of extraterrestrials having been very close to us for many years.

Meier has provided the world with not just hundreds of photos, but also metal samples, sound recordings, UFO imprints on grass and ice, videos, and several thousand pages of unique information published in over a dozen books of Contact Notes, and several other books he has authored. It is the most detailed and documented case concerning extraterrestrials visiting Earth and the crafts in which they fly.

The Meier case then is the most significant, important and longest standing UFO case, ongoing now for seven decades. It is understandable, therefore, that it is the case to have received the largest number of skeptical comments and reports critical of Meier's encyclopedic-like volume of evidence. These skeptical comments have extended to the WCUFO. At the outset, you too should be skeptical. A good scientist is skeptical because he or she requires *proof*; but at the same time, a scientist possesses an open and objective mind to find and accept the truth, whatever that truth may be, in this case concerning the WCUFO.

THE WCUFO

With its 43-odd brilliant spheres and assorted sophisticated parts of colored crystals and lenses, its various metallic pieces or accessories, golden features, and more, the WCUFO offers a huge amount of verifiable photographic evidence. The WCUFO is commonly called the "Wedding Cake UFO." Meier called it the "cake UFO," since its unique, *funny* cake-like appearance is so different from typical saucer shaped UFOs. Why does this WCUFO have so many spheres and odd parts that give it such a sophisticated appearance?

Could it be that other common flying disks (typical UFOs) have very similar interior details to this WCUFO, particulars that an outer cover hides from view? It is certainly a possibility. Perhaps this WCUFO is just a common UFO lacking an outer cover, hence its less aerodynamic, more complicated and unique appearance. It might be that the shape suffices for planned terrestrial activities only, while a protective cover is necessary for space flight.

Whether internally it is a common or unique UFO, this book proves the WCUFO authentic, so rather than asking whether the WCUFO is real or not future scientists and technicians are more likely to ask: How does it work? As scientists, they will seek the answer and experiment, and perhaps one day in futurity succeed in constructing a flying machine inspired by this design.

USEFUL BASIC SKILLS FOR PERFORMING THIS BOOK'S EXPERIMENTS

For the present, however, if you wish to research this strange UFO as a scientist or investigator, there are seven basic skills and abilities that in our opinion help or enable the smooth conduct of this book's eleven experiments.

In brief, these seven basic skills in order of appearance are:

1. Very basic photography abilities: take simple photos with a basic zoom camera or smartphone and upload them to a computer.

2. Very basic computing skills: perform simple computer image processing.

3. Curiosity, exacting observation and critical thinking. Anyone can easily follow this section by exercising critical observation and thinking.

4. Simple model-making manual tasks: cut and glue cardboard parts to make simple scale models.

5. Knowledge of junior high school math and trigonometry: at times conduct simple math calculations or use basic trigonometry.

6. Drawing basic scaled plans: make sketches and plans keeping scales and proportions accurate to scale.

7. Knowledge of some photography: understand depth of field, exposures, lenses, camera geometry, and image processing.Very basic computing skills: perform simple computer image processing.

Each skill has a specific representative graphic icon (see Figure 1), which appears in the text margin, informing what skill helps with the experiment. For convenient reference, you can also find the basic skills and icons at the very back of this book on page 157 under **Glossary of graphic symbols**.

Figure 1- Icons representing preferred basic skills to aid smooth conduct of the book's experiments.

Should you lack some or all of these required basic skills, seek assistance or work in a team with other young scientists or students who provide these abilities. Before beginning any of your experiments, therefore, check you have the skills, or assistance, and necessary materials to hand.

ADVICE ON PERFORMING THE EXPERIMENTS AND RESEARCH

We advise not to just rely on or trust any experts, scientists or investigators of the UFO phenomenon, including the authors of this book. Verifying the truth for oneself is entirely different from and preferable to accepting or believing somebody telling you what the truth is, no matter who that person is.

This book is a step by step guide through a process of discovering various aspects of the WCUFO. As you conduct your research and experiments, we suggest a neutral-positive attitude, which is a basic neutral stance with a positive frame of mind. Such an attitude is a healthy key ingredient for any budding scientist wanting to make a proper investigation and arrive at correct results.

The investigations and experiments in this guidebook have proven to be exceptionally revealing to us, so we invite you to investigate by performing some or all eleven of them personally. Trust in what you find yourself by simple actual scientific research, methods and testing. We anticipate you finding these experiments both fun and remarkably revealing and rewarding.

Chris and Francisco

October 2016

Chapter 1

Introduction to the WCUFO

Billy Meier, a Swiss farmer, claims he has been in contact with extraterrestrial beings (the Plejaren) that come from a place 80 light years beyond the Pleiades open star cluster, located in the Taurus constellation, in a different space-time configuration. Meier says they are human beings, like us, but with a higher level of technological and spiritual evolution. Meier took over a thousand photos that he claims were of their craft. Hundreds of the photos survive. Some show a strange UFO sporting three or four dozen reflecting spheres and various minor parts, like colored lenses and colored crystals. Meier took a few dozen photos of this particular UFO, which because of its peculiar form, has been called the "Wedding Cake UFO," or WCUFO in print. Meier refers to it as the "cake UFO."

Why is this WCUFO so complicated and why does it not look aerodynamic like a typical flying saucer? Maybe we can find a hint to the answer of this question in one of Meier's photos, of another UFO, called the "Plejaren miniscout" (Figure 2, bottom UFO), that sports three undercarriage spheres.

Figure 2- Billy Meier's photo showing a Plejaren beamship above and a miniscout below sporting three undercarriage spheres.

Do most of the "beamships," as the Plejaren are said to call their flying ships, have some form of these spheres? The miniscout shows three of them visible on its undercarriage, but in the WCUFO we see around 43 spheres. In other beamships, like the one above the miniscout in Figure 2, these spheres are not in evidence, perhaps because they withdrew inside the craft. Maybe the spheres are a type of small, sophisticated engine with some form of internal mechanism producing a magnetic field that interacts with Earth's magnetic field.

If this sphere magnetic field connection is correct, then it seems to us that having a few dozen spheres may make it more stable in our environment, or in other words, perhaps the WCUFO was designed for the Earth's magnetic field. Here is another noticeable difference between the WCUFO and other beamships Meier recorded on video: the WCUFO is very stable in our Earth's environment, hovering completely static, while the other beamships have a wobbling movement, as though they are floating on the Earth's magnetic field. Whatever, we think the WCUFO creators are probably showing us the inner workings of one of their ships. If so, this WCUFO is a UFO like the other beamships minus an outer cover.

Consider ourselves as future space travelers arriving on another inhabited planet. We want to give some ideas, not specific knowledge, to the inhabitants there that may help inspire them to progress and construct an automobile. So we construct an automobile capable of working in their environment and show it to them, but we remove the auto cover, so the core engine and mechanical parts are visible to their budding scientists. Perhaps, in a similar fashion, the WCUFO creators wanted us to learn through personal efforts the functions of the various parts and how to construct a similar ship.

It is possible that the WCUFO shows a technological missing link between our current vehicles and space traveling UFOs: a necessary technological step between our earthbound vehicles and vehicles that traverse the universe.

Initial designs of this WCUFO allegedly come from the 1920s suggesting this possibility. Contact Report #254 which Meier says he transcribed from his conversations with the ETs also implies this. Ptaah, an alleged extraterrestrial Plejaren human being, and friend of Billy Meier, explained to Meier the origin of the initial WCUFO designs in the following extract from Contact Report #254 including the original German text (28 November 1995; FOM 2015):

Ptaah:

4. We already worked with those flying devices, which you call the cake-ship, in the twenties, but it was indeed only at the end of the seventies that they were brought to the required status for their use on the Earth.

5. The form of these flying devices was specially thought up for the Earth, for which reason we made the effort to transmit the entire necessary specifications for the design to terrestrial scientists through impulse-telepathy so that, out of that, flying disks could be developed.

6. This impulse-telepathic information went predominantly to aerospace technicians, as I will designate these persons, whereby especially German engineers were included for this, to whom we transmitted exact plans for the external form as well as certain technical particulars which were responsible to transmit.

7. Thereby the German scientists also actually experimented, whereby they could construct halfway suitable flying disks, which according to our thinking at those times should have been used to constitute an air power through which an early-brought-about world peace should have been achieved.

8. However, the political machinations changed very quickly into a bellicose direction, for which reason we brought an end to further impulse-telepathic information to the German scientists and allowed the project to expire, whereby we however initially transmitted false information so that the flying disks could not be created specifically for warlike purposes.

Ptaah

4. Bereits in den zwanziger Jahren arbeiteten wir mit jenen Fluggeräten, die du als Tortenschiff bezeichnest, doch für den Einsatz auf der Erde wurden sie erst Ende der siebziger Jahre auf den erforderlichen Stand gebracht.

5. Die Form dieser Fluggeräte war speziell für die Erde gedacht, weshalb wir uns auch bemühten, impuls-telepathisch die gesamten notwendigen Angaben für die Form an irdische Wissenschaftler zu übermitteln, damit daraus Flugscheiben entwickelt werden konnten.

6. Diese impulstelepathischen Informationen gingen vorwiegend an Weltraumfahrttechniker, wie ich diese Personen bezeichnen will, wobei besonders deutsche Ingenieure dafür einbezogen wurden, denen wir genaue Aussenformpläne sowie gewisse technische Einzelheiten, die verantwortbar waren, übermittelten.

7. Damit experimentierten dann die deutschen Wissenschaftler auch tatsächlich, wodurch sie halbwegs taugliche Flugscheiben konstruieren konnten, die unserem Sinn gemäss damals dazu benutzt werden sollten, eine Luftkraft zu bilden, durch die ein frühzeitig herbeigeführter Weltfrieden erlangt werden sollte.

8. Die politischen Machenschaften jedoch veränderten sich sehr schnell in kriegerische Richtung, weshalb wir von weiteren impulstelepatischen Informationen an die deutschen Wissenschaftler absahen und das Projekt fallenliessen, wobei wir jedoch erstlich noch Falschinformationen übermittelten, damit die Flugscheiben nicht zweckgerichtet für kriegerische Zwecke erschaffen werden konnten.

According to this report, the original designs of these flying machines were for craft intended to operate on Earth, but unfortunately, due to the bellicose scientists and politicians we lost the opportunity to develop flying disks like these. While, as far as we know, our scientists have not to date been able to construct these, we do have remarkable photographs of them.

On October 22, 1980, Billy Meier saw and photographed a WCUFO hovering above his courtyard. Before this, he had been taking photos and videos of several flying disk type beamships. Now, Meier says the Plejaren brought the first of the WCUFOs to him: a 3.5-meter diameter flying ship. In Figure 3 we see an amazing photo of this WCUFO showing many colored crystals and small lenses, one of them, the blue one, inside or capping one of the spheres.

The house in the background on this cold day is Meier's home. In the sphere reflections, we can see a construction located behind Meier when he took this photo. This construction is the carriage house, a wooden building enclosing the parking lot (or courtyard) opposite Meier's home. We show later that Meier was located very close to the carriage house wall while this small 3.5-meter diameter WCUFO was flying in front of, and quite close to him. We also show that it was just less than 6 meters from his camera.

Figure 3- Photo #808 shows a WCUFO 3.5 meters in diameter with crystals, lenses, golden features and various little parts. Meier's home is behind.

Figure 4 is a wide angle view of Billy Meier's main property showing Meier's house on the right and the carriage house on the left. This WCUFO flew close to where the red car is visible in this photo. There were, however, no automobiles in the parking lot on the day Meier took his WCUFO pictures.

Figure 4- Wide angle view of Meier's main property in 1981.

Figure 5 shows the eleven photos Meier took this day. The picture numbers come from Meier's photo albums.

Later on, throughout almost one year, on numerous occasions, other WCUFOs are said to have come to Meier, and he was able to take additional pictures and a video of one of them.

Figure 5- Eleven photographs of the WCUFO taken by Billy Meier on October 22, 1980, in his front courtyard at 11:23 am.

Figure 6- Photo #829. March 26, 1981, at 6:19 am. A 7-meter diameter WCUFO.

On March 26, 1981, Meier took several photos of a 7-meter WCUFO between him and his green trailer. Only two came out successfully (Figure 6). The front of the WCUFO is just out of focus, and its middle, the trailer, and the horizon are clearly in focus.

This depth of field (DoF) is very revealing. Using a 55 mm lens like Meier's Ricoh and ensuring the horizon is in focus on an overcast day things go out of focus at around 5 -13 m from the camera (on f5.6 – f11). So, if the Ricoh was used the WCUFO is 5 - 13 m away. With a 55 mm lens, however, it is impossible for the WCUFO to occupy 2/3 of the photo width, as it does here, at less than 10 m away; and beyond 14 m the entire WCUFO is in focus. Only one possibility remains for the Ricoh: f5.6 giving a DoF of 14 m – 3.9 m. At 13 m away a 7-meter craft may fit the picture width, but Meier was unlikely so far from the craft and so unlikely to be using his Ricoh. Indeed, Meier says he was about 6 m away using his Olympus 35 ECR 42 mm lens despite recently having acquired his Ricoh.

The Olympus is auto set, probably the usual f8 or f11, with a DoF of 7.4 m to infinity. So the unfocused WCUFO front face is 6 - 7 m away and the horizon in focus, agreeing perfectly with the photo details and Meier's claim. The center of the WCUFO is then about 10 m away correctly occupying 2/3 of the picture's width. So, all the photo details suggest this shot was taken with Meier's Olympus 35 ECR from about 6 m away. Furthermore, we cannot be looking at a small model. With either camera, due to their DoFs, a small 500 mm model one meter away with its front out of focus and middle and back in focus, as the photo shows, results in the distant horizon out of focus – which it is not. So this photograph shows a large object, and Meier's 7-meter-diameter claim exactly fits the photo data.

Figure 7-   Photo #834. April 3, 1981, at 1:10 pm.

Figure 8-   Photo series among the treetops.

So we would not expect to see any noticeable shadow. Incidentally, increasing the brightness of Photo #829 (Figure 6) shows its left side from our viewpoint *behind* tree branches.

A week later on April 3, 1981, Meier says he was flying among the treetops at around 1:10 pm on a 7-meter WCUFO taking photos of a companion 3.5 m WCUFO. He took several pictures of this WCUFO behind tree branches. Figures 7 and 8.

Meier says Quetzal, a Plejaren commanding these ships, accidentally broke off the upper portion of one tree, a photo of which exists (*They Are Here*).

Analyses of this WCUFO's sphere reflections prove it is a sizeable object, not a small scale model. A scale model UFO would have to be very close to the tree at front; and the tree reflection in the spheres would then cover a significant area of the spheres, but it does not.

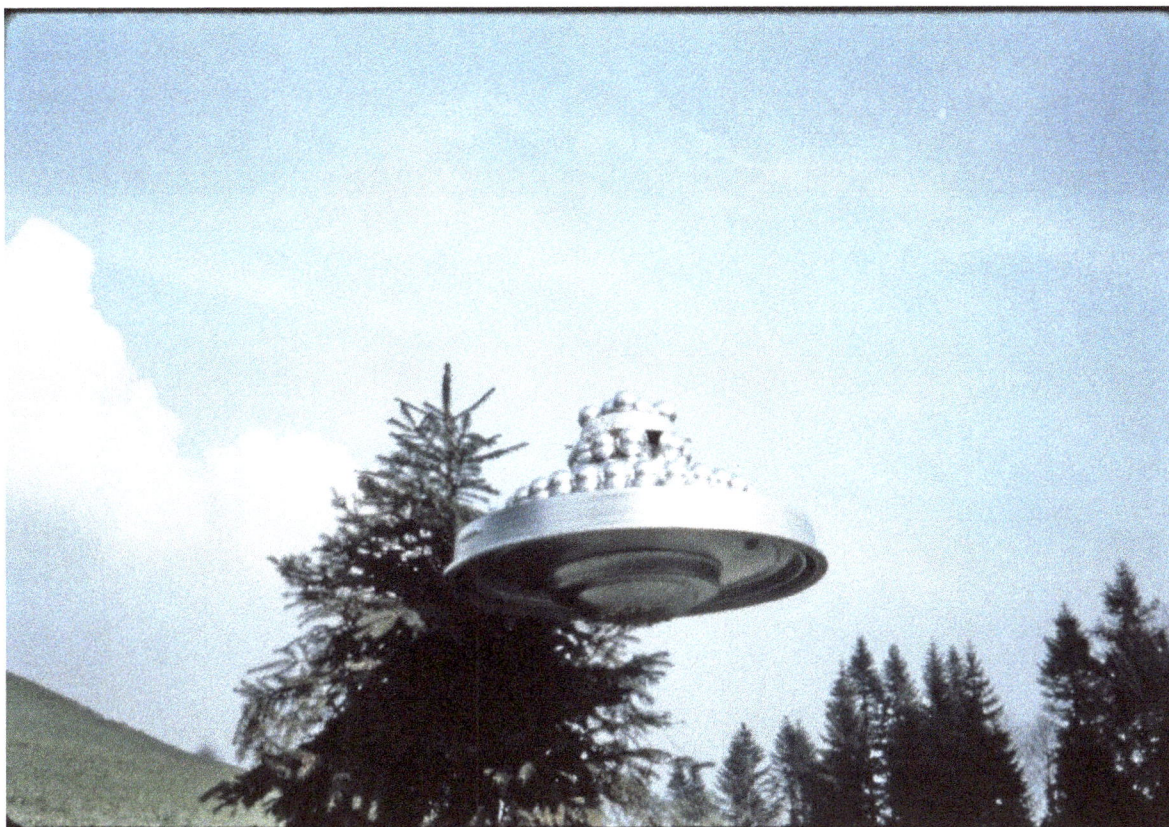

Figure 9-   Photo #844. April 3, 1981, at 1:35 pm. WCUFO casting a huge shadow over a young Norway Spruce.

A few minutes later the 3.5-meter WCUFO stayed motionless immediately in front of a 6 to 8-meter tall Norway Spruce tree. Meier, walking towards it on the ground, took a series of photos. Figure 9 shows a shot from the series. With "brightness" enhanced it clearly reveals a huge shadow the craft casts over the tree's sizeable branches. (For better views see Experiment 9 exercise 3 and *They are Here*). Notice also the large pine needles, and the dark green reflection on the left of the WCUFO undercarriage confirming its proximity to the large spruce. Calculations shown later, based on the camera Meier used, put his location at around 16 meters from the craft (see Figure 67).

One hour later at 2:33 pm the same day, Meier took another series of photos, and a video, of this WCUFO close to another tree. Figure 10 (top) shows a shot from the series. With image processing (Figure 10, bottom) the branch and leaf structure of the tree in front of the WCUFO become clearly visible.

Figure 10- Top: Photo #850. April 3, 1981, at 2:33 pm. Bottom: The same photo image processed to enhance this Norway Spruce tree in front of the WCUFO with its bright red crystals.

On the night of August 2, 1981, Meier says the WCUFO returned. The pure silver color now looks golden at night. Studies of its bright sphere "reflections" suggest that rather than reflecting street lights it *emits* light. Figure 11 shows the 7-meter WCUFO behind a Mercedes-Benz vehicle, and a little tree closer to the camera. Because this is a night photo the tree is totally out of focus due to the camera depth of field effect and the car is a bit blurry. The light trace on the top left is said to have been caused by a small telemetric disk moving very fast during the photo exposure. Telemetric discs, according to Meier, are little flying monitoring devices the Plejaren say they use.

Figure 11- Photo #999. August 2, 1981, at 2:18 am.

Meier took several WCUFO photographs in this location very close to the parked car.

A few days later on August 5, again at night, Meier says he boarded a 3.5-meter WCUFO and flew away hovering above central Switzerland, with another 7-meter diameter WCUFO in front of the one he occupied. He took what many think is the most remarkable picture of the WCUFO: Photo #873 (Figure 12) showing a WCUFO with its central core extended upwards by about 13 centimeters. Enhancing the brightness of this picture renders visible the terrain and a fence pole below the WCUFO. Enhancing this WCUFO image reveals a strange violet or wine red halo around it, and a foreground landscape beneath it. Remarkably these details had remained undiscovered for over 30 years. Furthermore, a detailed analysis of this picture suggests this WCUFO has the capability to hide the light it emits by a screening field around it.

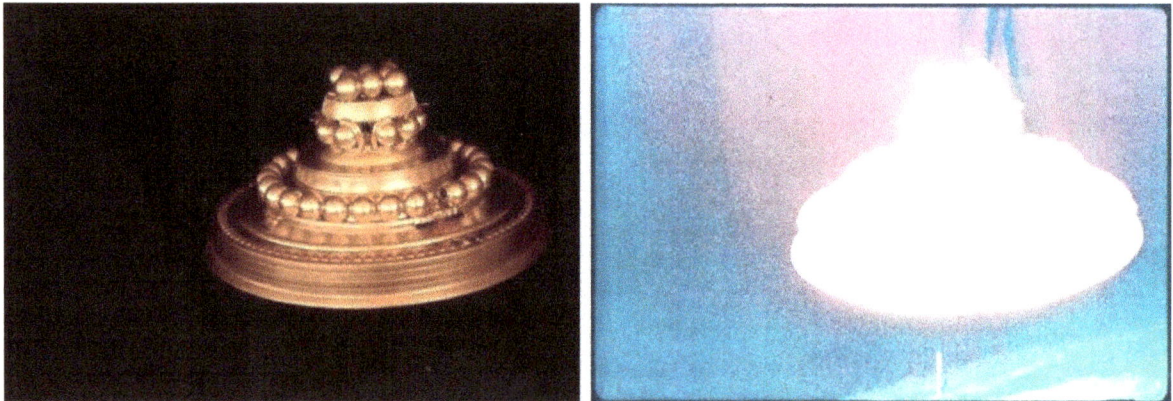

Figure 12-   Photo #873. August 5, 1981, at 2:48 am.

More photographs of the WCUFO are available in the book *Photo-Inventarium*. We also suggest readers review the authors' upcoming book *They are Here: Compelling Billy Meier evidence of real extraterrestrial ships present on Earth*, which details some intriguing aspects of this machine.

Anyone can find similarities between the parts of this WCUFO and common household items. The base of the WCUFO looks like an available rubber or plastic trashcan lid, and there are additional parts similar to household items. It seems that after this WCUFO was photographed, for reasons unclear, some people were making different everyday items very similar in scale to a 55-centimeter diameter model. Not surprisingly, therefore, would-be debunkers support a controversial idea that the WCUFO is a small model and so the WCUFO case a hoax. However, in this book we establish that this and other WCUFOs are large objects, of at least two sizes: approximately 7 meters and 3.5 meters in diameter. Performing the eleven experiments in this book proves these findings to be true, and with them, any budding young scientist can find the real size of this flying machine and understand why it cannot be a small 55-centimeter diameter model.

Chapter 2

Introduction to the WCUFO photo experiments

In this chapter, we introduce and detail this book's eleven WCUFO experiments. Each subsequent chapter covers a different experiment and most involve the WCUFO spheres and their reflections; but what can the sphere reflections of this WCUFO tell us?

The spheres work as mirrors, showing what is in front of, and around the craft. When we look at ourselves in a mirror, we see not only our face but also a few objects proximal to us and others in the background behind us. Moving away from the mirror our face reflection becomes smaller, and conversely moving closer to the mirror, our face reflection becomes bigger. The same effect happens with the spheres on the WCUFO, but with the difference that the mirror now is spherical. Spheres make the reflected objects look smaller and farther away, but they also display the surrounding environment in a panoramic image of more than 200 degrees. This view is much wider than that of a camera wide angle lens.

So the reflections on the WCUFO spheres give a unique opportunity to see images of several objects around the WCUFO that inform us how far away the objects are, no matter the reflections often lack the sharpness we would prefer. On the WCUFO hovering above Meier's courtyard parking lot the spheres reflect a sizeable wood construction, which is the carriage house. Knowing the size of this carriage house, which we do, and comparing it with the image seen in the spheres, we can know how far away this WCUFO is from the camera, and also how big it is. Doing the calculations is simpler than it sounds, and performing the suggested experiments should guide you comfortably to personal conclusions. So, this is your invitation to research as a scientist and enjoy making and charting your findings.

The main finding and conclusion of the original *Analysis of the Wedding Cake UFO* were that no matter this UFO photographed by Billy Meier in the 1980s might look like a toy made of household items it is proved to be a sizable object more than 3 meters in diameter, and close to 3.5 meters. In the experiments that follow, you find simple ways to experiment with some local environmental features around the mysterious WCUFO and discover in a practical manner its actual size, and some of its interesting characteristics that confirm the findings.

The experiments mostly follow a regular order based on results and verified data of previous experiments. Teachers assisting students with these chapters in classes may, however, decide to do any number of them in a different order depending on curriculum requirements, interest, aptitude, costs, facilities, availability of materials, or class time constraints. The final three, Experiments 9, 10 and 11 can stand alone and could be inserted at any place in a study course deemed appropriate regarding variety, interest, cost, or practicality; or they could be saved for use at the end of the semester if time is available.

The armchair reader can enjoy just to read and follow the experiments, checking everything mentally step by step to each conclusion.

List of experiments

1- **Understanding spherical reflections**: This practical experiment helps you understand the nature of sphere reflections, rules that govern them, and their significance for the WCUFO. (Chapter 3.)

2- **Making a plan view of Billy Meier's property and site,** the photographed location: This online experiment puts you approximately in Meier's location and shows the buildings around him. (Chapter 4.)

3- **Establishing Meier's location**: This observation and critical thinking experiment enables knowledge of Meier's exact location on his property. His location is imperative because the size of the WCUFO depends on its distance to him and his camera. (Chapter 5.)

4- **Making and using a scale model of Meier's residential area:** This practical indoor model puts you in Meier's shoes as you replicate taking his photos. It also enables experimentation with various WCUFO sizes and configurations. (Chapter 6.)

5- **Making and using big scale models of Meier's site & camera:** With these outdoor models you experience Meier's exact physical relation to the WCUFO and its environment, and view scenes just as his camera did. This field experiment requires an open space and outdoor resources. (Chapter 7.)

6- **Calculating the WCUFO angle of view and size:** Here you learn a graphic method to calculate the WCUFO *angle of view* from any WCUFO photo. The angle of view then enables a good approximation of the WCUFO size. (Chapter 8.)

7- **Distance estimation using the camera formula:** The *camera formula* relates to camera geometry and enables good estimates of distances to objects once we know their real size. Having established the approximate WCUFO size in experiment six, the camera formula now enables a simple investigation to determine the distance between the WCUFO and the camera. (Chapter 9.)

8- **Mapping the WCUFO's local environs:** A simple graphic method of mapping in detail the WCUFO's local environmental surroundings by charting the position of the spheres' reflected images. (Chapter 10.)

9- **Image processing photo images to discover hidden details:** In this experiment, you analyze available WCUFO photo pictures and reveal previously hidden details through image processing. (Chapter 11)

10- **Making a stereoscope for viewing 3D WCUFO pictures:** This simple, practical and enjoyable experiment shows how 3D images work, how to see the WCUFO in 3D, and how to confirm WCUFO sphere reflection details in 3D. (Chapter 12.)

11- **Height changing capability of a WCUFO:** This basic geometric analysis confirms whether one of the WCUFO photos shows a WCUFO capability to extend its central core upwards. (Chapter 13.)

Chapter 3

Experiment 1: Understanding spherical reflections

INTRODUCTION

This experiment helps us understand the nature of sphere reflections, rules governing them, and ultimately what they tell us about the WCUFO. You confirm three critical key sphere reflection rules to understand not just the WCUFO sphere reflections but the WCUFO itself.

MATERIALS

First, find some highly reflecting spheres, like Christmas tree balls or bigger, and experiment looking at reflections in them. You could set up an arrangement like Figure 13. Studying the reflections in your spheres reveals the following three reflection rules:

Sphere Reflection Rules

Rule a (i) - *The size of the reflected image is inversely proportional to the object's distance from the sphere center.*

Increasing the distance reduces the size of the reflected image.

Rule a (ii) - *Wherever the observer locates on a line between an object and the sphere, the reflected images of other stationary objects remain the same size.*

Rule b - *The observer's eye or camera lens reflected on a spherical surface is always at the center of the image.*

Rule c - *No matter what size the reflecting sphere is, the size of the reflected object is always in the same proportion to the sphere's size, given the same distances between the sphere and the object.*

Being purely scientific we could say: *The absolute magnitude of the reflected image is directly proportional to the sphere diameter.*

A reflected object always occupies the same proportion of the sphere. If your body covers 50% of the sphere diameter, it is always 50% in any size sphere at the same distance from your body.

An exception: These rules always stand for small to medium size spheres. When using enormous spheres with a radius very similar to or larger than the distance between the sphere's surface and the observer, there are variations in these rules. For example, for a 5-meter radius sphere with an observer very close to or beneath its surface.

METHOD

Testing and confirming the three sphere reflection rules

Confirming *Rule a (i) and Rule a (ii)*

We tested with spheres of four different sizes: 200 mm, 100 mm, 48 mm (blue sphere in Figures 13 and 14), and 38 mm in diameter. Figure 13 shows their arrangement hanging from the wall. The observer with his camera (an iPhone) stood in front of the arrangement at varying distances from the sphere centers. An area of the floor was covered with a white blanket, which, with the white living room walls and ceiling provided an excellent contrast to leave the observer's body reflection clearly visible. He took photos at various distances to the spheres and measured the distances on the floor.

Now, do the same yourself to learn and become acquainted with how sphere reflections work.

We took zoomed pictures at staggered distances of 1 m to 4 m of the arrangement in Figure 14. The first column shows 1 meter from the camera to the sphere center and the fourth column from 4 meters away. The top row shows zoomed out photos of all the spheres, and the bottom row zooms in on the biggest one, the central sphere.

Since our spheres were close to a wall, its reflection produces an encircling white ring giving the appearance of a sphere within a sphere (Figure 14). This effect is due to the previously mentioned 200-degree-plus view that the sphere provides. Doing this experiment outdoors, without a wall close to the spheres, eliminates this effect. Just ignore it if experimenting indoors, being sure to note the bigger sphere contour, the red circle in Figure 14.

Figure 13- An arrangement of four spheres hung on a wall.

Yellow marks in Figure 14 close to the center show the photographer's full body size. The percentages under the pictures note the ratio of the size of the object's reflection (the photographer's body) to the sphere's size at the four different distances. For ease, we calculated percentages from magnified images on the computer monitor. For example, the sphere (red circle diameter) located at 1 meter measured 185 millimeters on the monitor while the photographer's body was 48 millimeters tall. Dividing 48 by 185 equals 0.26 or 26%.

Each 1 m closer to the sphere, the reflection size increased by 144%, and each 1 m more distant from the sphere the reflection size reduced by 31%. So in our case, the reflected image depended on or was approximately inversely proportional to the object's distance from the spheres. Our tests suggest a linear relationship between the reflection size and the object's distance from the sphere.

We have now confirmed **Rule a (i)**:

The size of the reflected image is inversely proportional to the object's distance from the sphere center.

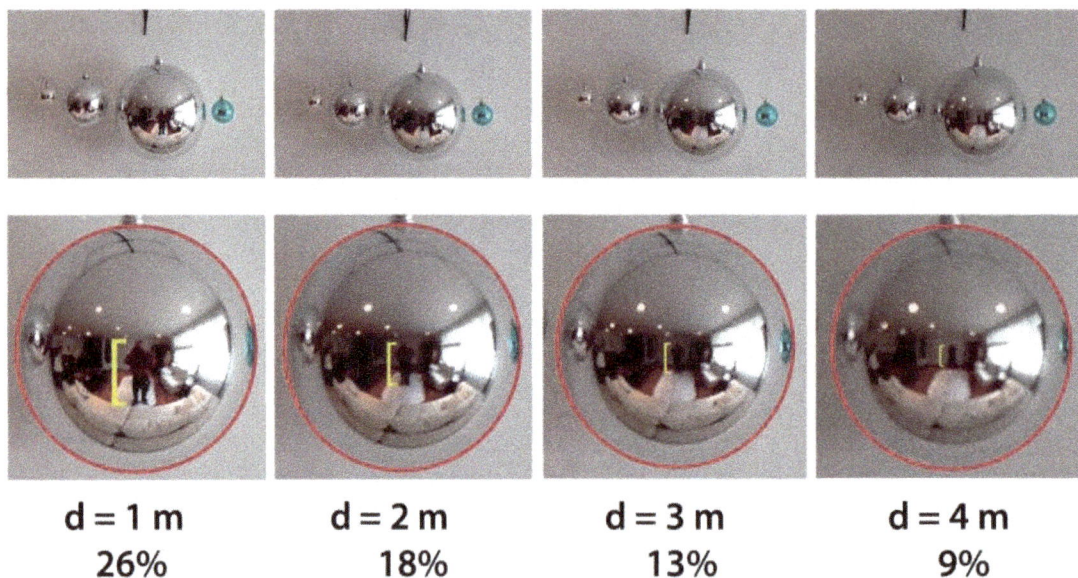

| d = 1 m | d = 2 m | d = 3 m | d = 4 m |
| 26% | 18% | 13% | 9% |

Figure 14- Yellow marks denote full body reflection sizes of the photographer on the spheres (red circles) at different distances.

In this test, the object was the photographer's body, which looks bigger when closer to the sphere, and smaller when farther away. Note that all other objects look the same size since their distances from the sphere have not changed. Only the object that moves closer or farther away from the sphere changes in size.

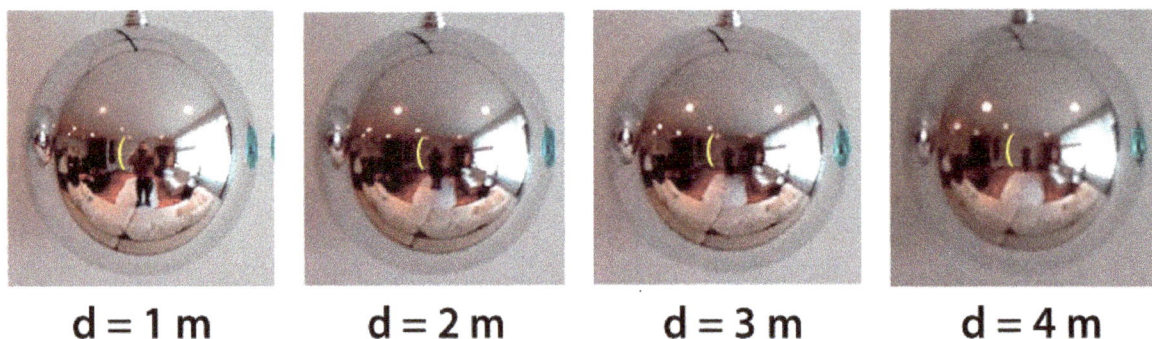

| d = 1 m | d = 2 m | d = 3 m | d = 4 m |

Figure 15- Back wall reflections. All stationary objects remain the same size irrespective of the observer's distance. Distances "**d**" are from the camera to sphere center.

To confirm **Rule a (ii)** we vary the photographer or observer's position between the sphere and another object. Figure 15 shows the pictures taken on the bigger sphere at different distances. The reflected object considered now is the back wall. The camera, again, was located at 1 meter, 2 meters, 3 meters and 4 meters. All spheres are the same size.

We noticed previously that the body or object size becomes smaller as it moves away from the sphere, but the size of the back wall (Yellow marks in Figure 15) remains the same. All objects in the living room remain the same size, no matter how far away the photographer takes the photo. The observer could be farther away from the sphere than the reflected object, beyond the back wall, and the rule would stand.

We have now confirmed **Rule a (ii)** :

Wherever the observer locates on a line between an object and the sphere, the reflected images of other stationary objects remain the same size.

Rule a (i) and **Rule a (ii)** are crucial in determining the actual size of the WCUFO. As it hovered above Meier's parking lot, in its sphere reflections, we see the carriage house that Meier was very close to when he took these WCUFO pictures. The size of this reflection tells us how far away the carriage house was from the spheres and the WCUFO.

We found, and show later, that the nearest WCUFO sphere for Photo #800 was around 6.2 meters from the carriage house wall or Meier's camera proximal to the wall. We confirm the proximity shortly. So, whether the WCUFO was about 3.5 meters in diameter or a small scale model of half a meter in diameter made from a trashcan lid, the nearest sphere can only be at a set distance from the carriage house wall of around 6.2 meters. (See Experiments 4, 5 and 6.) If Meier had photographed a small model at 6 meters from the camera, it would have looked tiny in the photos and could not cover the large area of the picture that it does. We return to this point in later experiments.

Confirming *Rule b*

To confirm **Rule b** walk around your spheres and check where your eyes appear in each of the reflections. In our experiment, the photographer used the same arrangement used to confirm ***Rule a (i)*** and ***Rule a (ii)***, moving to the left, the right, and below. See Figure 16 for the results.

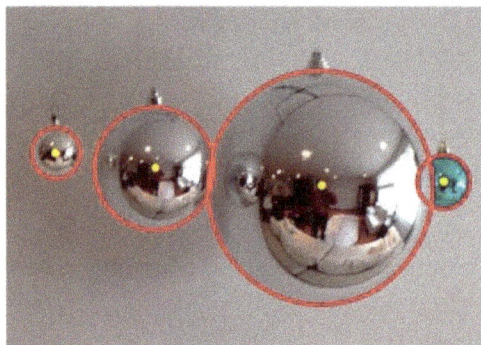

From the left.

From the right.

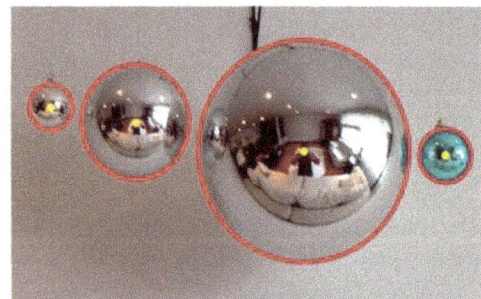

From below.

Figure 16- Yellow spots show the position of the camera lens reflection in each sphere (red circle). Photos taken from the left, right and below.

Figure 16 shows a yellow spot drawn at the precise camera location on each reflected image, and a red circle drawn around each sphere. The yellow spots are always at the center of the red circles. Neither the location of the observer or camera matter nor how big the spheres are: if they are true spheres, the camera lens or eye is always at the center.

So we have confirmed **Rule b**:

The observer's eye or camera lens reflected on a spherical surface is always at the center of the image.

Look at several WCUFO photos taken by Meier. In every one, the location of his camera reflection is always at the center of the sphere. So, if we see a few trees at the center of the reflection, we know Meier's camera was there at the center, inside those trees.

Confirming *Rule c*

Rule c concerns sphere size. If Meier made a little model using small Christmas tree balls, the spheres would have been tiny, close to one inch or 25 mm in diameter. However, if this WCUFO was the size of a little car of about 3.5 m, the spheres might be around 12 inches or 300 mm in diameter. So, to conduct tests, how big must the spheres be? The answer is: it does not matter. Every size of reflecting sphere produces the same reflection result. The reflection of the carriage house behind Meier when he took his pictures occupies the same proportion of any size sphere, given no change in distance between the WCUFO and the carriage house. If the width of the reflected carriage house image covers 30% of the full sphere diameter, it covers 30% on any size sphere, given the same location and distance. The percentage depends on the distance between the object reflected and the sphere center, **Rule a (i)**, not on the sphere size, which is irrelevant. (Again, as long as the spheres are small to medium sized, and not enormous.)

When confirming **Rule c**, the sphere images used were all taken at a camera distance of 1 meter from the spheres (Figure 17). Zooming the images on the computer screen, we measured the sphere diameters and lengths of the photographer's full body reflection. The body heights were divided by the sphere diameters and expressed as percentage values. (See Figure 17).

Given a 1% margin of error, the percentages are just about the same. The 1% difference is due to the difficulty in measuring distances precisely for the small spheres since their images are a bit blurred and the measurements tiny. We see that whatever size the reflecting sphere is, the reflected object always occupies the same proportion of the sphere.

Page 23

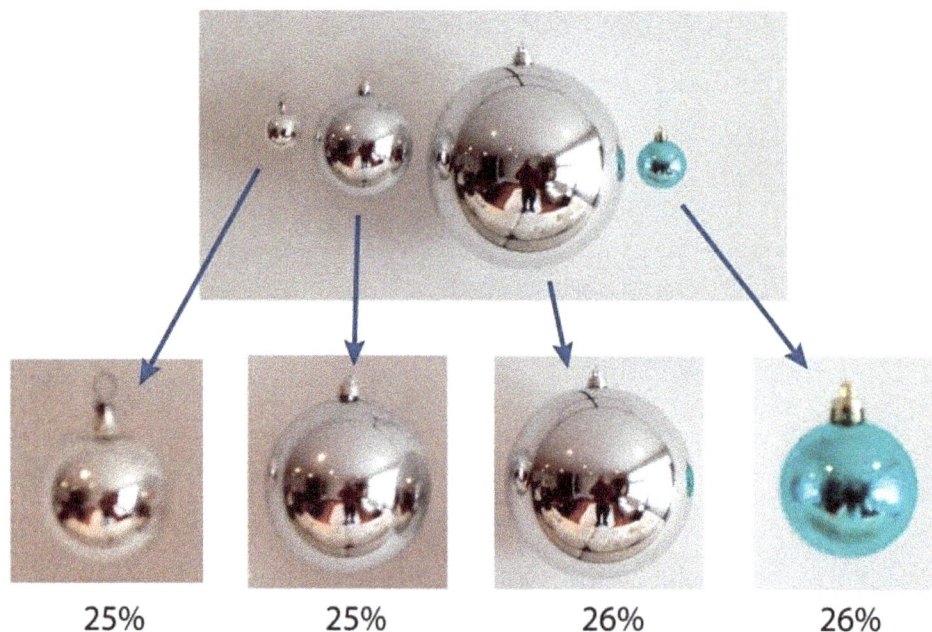

| 25% | 25% | 26% | 26% |

Figure 17- The full body height divided by the sphere diameter expressed as a percentage for each sphere.

We have now confirmed **Rule c**:

No matter what size the reflecting sphere is, the size of the reflected object is always in the same proportion to the sphere's size, given the same distances between the sphere and the object.

Alternatively: *The absolute magnitude of the reflected object is directly proportional to the sphere's diameter.*

These rules always stand, as previously noted, for spheres with a radius less than the distance between the observer and the sphere's surface.

A large reflecting sphere (300–600 mm) is best for performing several tests since it produces clearer images and therefore greater accuracy in measurements when zooming the image.

Chapter 4

Experiment 2: Making a plan view
of Billy Meier's property and site

INTRODUCTION

Now that we know how reflections on the WCUFO spheres work, we can establish a good approximation of Billy Meier's location on October 22, 1980, when he took the 11 pictures of this WCUFO in front of his house (Figure 5).

Unfortunately, we cannot all travel to the Swiss mountains and visit the Semjase Silver Star Center (SSSC), a place of unusually high UFO activity in the 1970s and 80s and where Meier had been living for some years. An excellent option for those unable to visit the SSSC is to make a map of Meier's property from available photos. Though not high precision, such a map can still give an excellent working estimate of WCUFO details and Meier's location. It can put us very close to where Meier was standing, and show the buildings around him.

MATERIALS

All you need to conduct this experiment is access to Google Earth maps, some translucent paper and a pen or marker to trace outlines of Meier's main buildings from the map on your computer screen.

METHOD

Start with Google Earth. Open it and in the upper left area "search" field write the latitude and longitude of the SSSC: 47° 25' 00" N and 8° 54' 23" E (or 47.41666 and 8.90639). These readings take you to the latest satellite maps of Meier's property. In 2013 we obtained the image in Figure 19 (left side). Currently, in 2015, there is a higher resolution image, but shadows may hide some details (Figure 18). To compare both together see Figure 19. Press the down arrow just below the N sign (North) on the navigation controls to ensure you are in "top view."

In Figures 18 and 19 (right side), the bigger construction is Meier's house and the smaller one at bottom left is the carriage house. Meier was located very close to the carriage house looking at his home. The WCUFO hovered close to the central orange mark. The eleven photos taken (Figure 5) show Meier's house in the background. Figure 4 gives a wide-angle view of the area from a nearby hill, with Meier's house on the right, and the carriage house on the left. Noticeably, several trees have grown considerably since 1980.

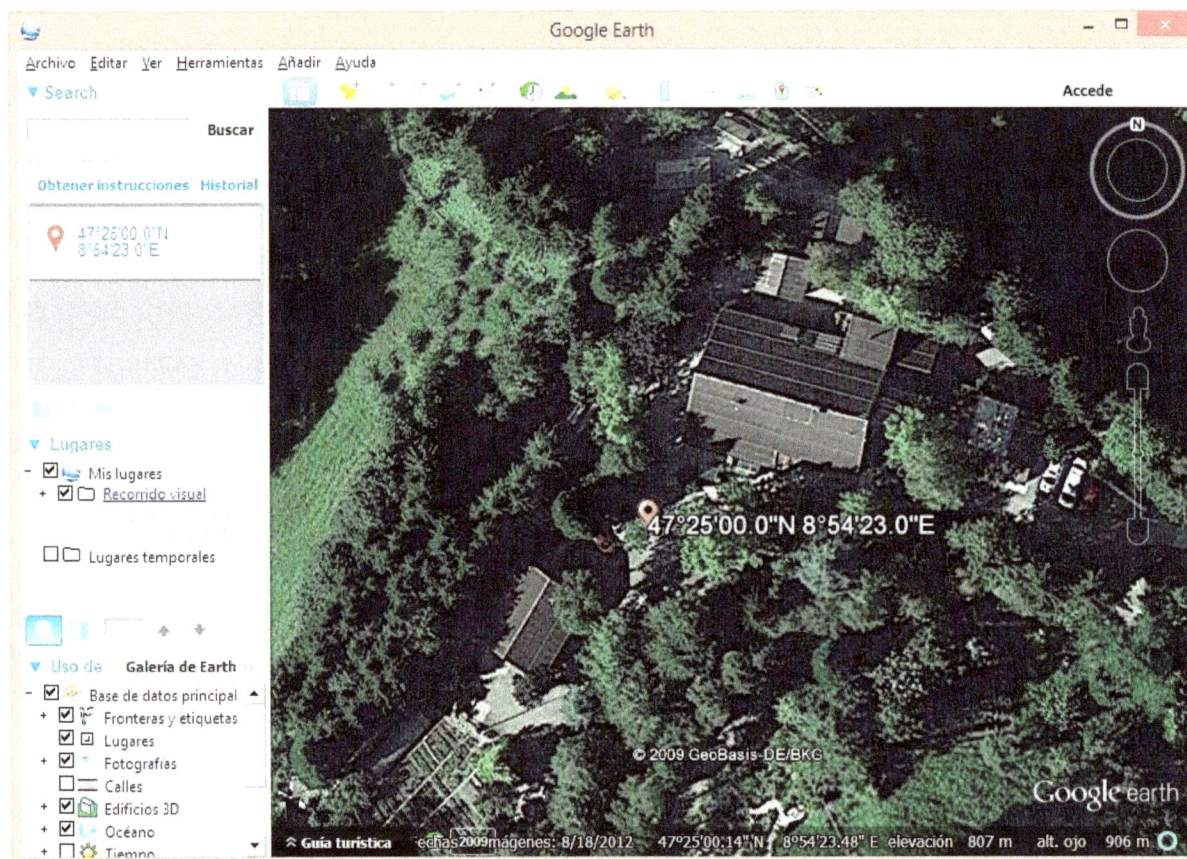

Figure 18- Google Earth map of the SSSC in Jan. 2015.

With Google Earth still open, using the ruler tool measure the length of the fully exposed half of the carriage house roof, as in Figure 19, right side. We obtained 5.5 meters. The whole roof is twice this value: 11 meters. Christian Frehner measured this length at the SSSC and arrived at 10.3 meters (Annex B Figure B1). So there is a small difference of 70 centimeters between our two measurements, perhaps due to the satellite angle, roof pitch or overhang, but it is sufficient to estimate the WCUFO size.

Below we outline two methods for drawing a map of this image, but first:

In Google Earth draw a line measuring 10 meters using the ruler tool. Then, without closing the ruler window, select the menu option "Save" and "save image" to save your JPEG Google Earth image file. Remember to ensure a "top view" by pressing the down arrow below the N symbol (North) in the navigation arrows on the right.

Figure 19- Google Earth map details of the SSSC. Left 2013 image and right 2015 image. Half the carriage house roof measures 5.5 meters.

First method for drawing the map: Open your saved image or use the screen image. Place a piece of translucent paper over the computer screen. Then on the paper draw a tracing of the carriage house, Meier's house, and the 10 meters mark you made on the map that is now on the screen.

Second method for drawing the map: use a computer tool, like "paint," to draw these constructions on the saved image. The resulting image we obtained with "paint tool" is in Figure 20.

This map is accurate enough for the following experiments that determine the position and size of the WCUFO. Annex B provides more detailed information on carriage house dimensions and a similar but more detailed plan view. For the following experiments, the plan view obtained with either the first or second method (Figure 20) will be fine.

Figure 20- Construction layouts drawn with Paint Tool on the Google Earth image. The 10-meter yellow line marking is useful later.

Chapter 5

Experiment 3: Establishing Meier's location

INTRODUCTION

Establishing the exact location of Meier's camera while he was taking these photos is imperative. The size of the WCUFO in the photos depends on how far away it was from his camera. We find in a later experiment that the nearest WCUFO sphere was a little more than 6 meters away from the carriage house wall. The WCUFO in these photos covers almost the full width of the picture, so either Meier is using a small scale model at just one meter away, or he is shooting a big object from much farther away close to the carriage house wall. If he is using a small model, he must be at a distance of 5 meters from the carriage house wall, towards the center of the parking lot, not close to the wall. If we find Meier's location close to the carriage house, it means he was photographing a significantly sized object, not a small model.

This experiment also cites findings from an earlier experiment by Christian Frehner.

No special physical materials are necessary for this experiment other than your critical faculties for asking and answering important questions. You exercise these faculties in this experiment while studying some of the WCUFO photos, and responding to questions about them.

METHOD

Look at and study for a moment the following images in Figures 21 and 22:

Figure 21- Overlay of photo images #799 (left), and #800 (right) showing Meier's entire house in the background with its left gable end wall defined by red lines.

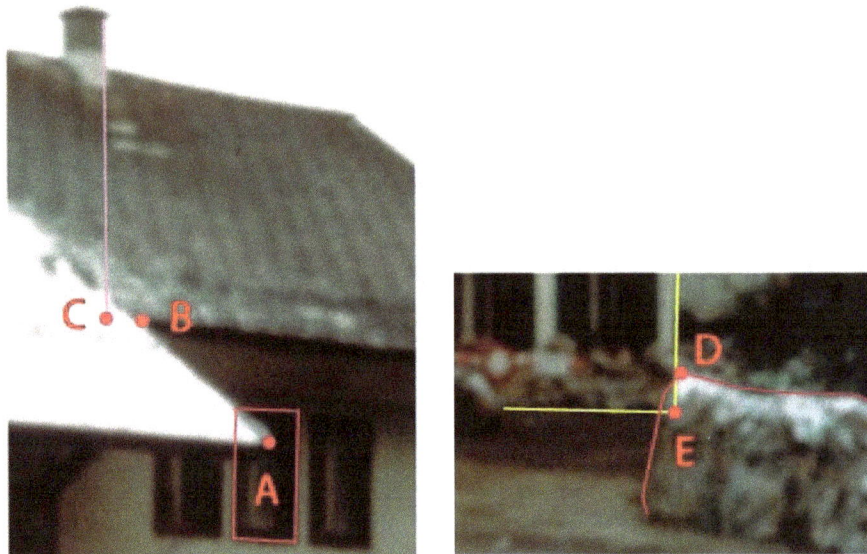

Figure 22- Alignments A, B, C, D, and E pinpointed by red dots: Dots A and B are roof corners; C is the projection of the right side of the chimney; D is the top of the little stone wall; E is the lower corner of the house.

Questions to answer on some of the photos come from the following 12 enumerated points. Should you experience difficulty answering the questions, refer to the three Sphere Reflection Rules that you previously proved (page 17). Later you review your answers by comparing them with our own.

1. Figure 22, left side, shows a red rectangle outlining the second window from left on the second floor of Meier's home. Point A is at the extreme roof corner. Are point A and the window equidistant from the camera? If not, what do you estimate the distance is from the window to point A on the roof? (You can use the detailed map in Annex B, Figure B6.)

2. Figure 22, right side, shows point E at the lower corner of Meier's house, and point D at the crest of the little stone wall. Are points D and E equidistant from the camera? What do you estimate the distance is between them? (Again, you can use the detailed map in Annex B, Figure B6.)

3. Look at the miniature Photos #799, #800, #803 and #808 in Figure 5. Are points A, B, C, D, and E in the same position?

4. If Meier and his camera, moved one meter to the right or left, between taking these photographs, would all or any of points A, B, C, D, and E have remained in the same position? What does the fact that they stay in the same places in Meier's photos tell you?

5. Would taking this WCUFO photo session with a different camera to Meier's roll film camera, like an electronic camera, or using a different lens, like a wide-angle or telephoto lens, have changed any alignments of points A, B, C, D, and E?

Referencing an earlier experiment

Christian Frehner performed an experiment some years ago with a large reflecting sphere at the SSSC. After studying Meier's photos, he attempted to locate himself at the exact position in the front yard where Meier took his pictures. Figures 23–26 show what Frehner's photos captured.

Figure 23 shows a photo taken with an electronic camera. Notice there is more vegetation, and Meier's house has had a few renovations. The camera did not use a high resolution (or lacked the capability), so these sphere reflections are not clear. Frehner's other photos at closer distances (Figures 24 and 25) display clearer details on the reflecting sphere. This first shot from Frehner, however, helps us check his alignment and to compare it with Meier's photo. We did our best to define points A, B, C, D, and E in Figure 23.

Figure 23- A Christian Frehner photo experiment of a few years ago. Alignments A, B, C, D, and E pinpointed by red dots as in Figure 22.

Two more questions:

6. In Figure 23, left: How does Meier's left wall elevation red outline here compare with the red outline in Figure 21?

7. Do points A, B, C, D, and E in Figure 23 have the same alignments as in Meier's WCUFO photos? What does your answer reveal?

In Frehner's experiments, he photographed the reflecting sphere held by a volunteer at just 2 meters and 1 meter from the camera while remaining in the same location as in Figure 23. See Figures 24 and 25.

8. Do you see the photographer's image reflected in the sphere in Figures 24 and 25? What about the carriage house?

9. How far away do you estimate the carriage house wall to be from the photographer? (See also Figure 26.)

Figure 24- Reflecting sphere at 2 m from the camera, taken from a position close to the carriage house wall looking north. Right: Zoomed image.

Figure 25- Reflecting sphere at 1m from the camera showing the photographer at the center. Right: Zoomed image.

10. This sphere is about 30 centimeters in diameter. If Frehner had used a smaller sphere, like the 2.5 cm ones on WCUFO scale models, would the corresponding area of the reflected carriage house wall be different on the smaller sphere? (Again, if unsure refer to the proven reflection rules on page 17.)

11. If Frehner and his camera had moved forward five to ten meters towards the right edge of Meier's house would there have been changes in the alignments of points A, B, C, D, and E? What about the left end wall profile of Meier's house? (See your map made in Experiment 2 or the Annex B map, Figure B6.)

12. What are your conclusions?

Answer questions 1 to 12 and compare your answers with ours.

Figure 26- Carriage house from the little stone wall. Point F marks Frehner's location during his experiment. Point M marks Meier's probable location very close to Frehner's. In 1980 there was no garden bench.

Our answers are:

1: No. Point A and the window are not equidistant from the camera. The window is approximately 5 meters beyond the roof corner.

2: No, they are not equidistant. The little wall of stones is much closer. We estimate 19 meters closer. It could be a bit different since Google Earth does not show this small wall. We estimated the wall's position from Figures 4 and 26 and the detailed map of Figure B6 (Annex B).

3: Yes, they are. The alignments are the same where visible. These WCUFO photos in Meier's courtyard indicate that he did not move, at least not in the photos where alignments are visible.

4: If Meier had moved, the relative positions would have changed. One meter is enough to appreciate a shift, especially in the relationship between the little stone wall and the house in the background, since around 15 meters separates them. In other words, the visible left part of this wall is half the distance between the camera and the right edge of Meier's house. Since these alignments do not change in Meier's Photos #799, #800, #803 and #808, it tells us Meier was in the same place while catching these photos.

5: Alignments do not change when using different lenses or cameras. Even viewed with the naked eye the alignments remain the same. The alignments are because the rays of light from these points come from the same direction. To prove this yourself, photograph, or just focus on, two objects with a through-the-lens camera, one object farther away from the other. If they look to be in the same position, without changing the camera location, modify the lenses or use the zoom, and you see they remain aligned. This fact proves that calculating Meier's location in his front yard bears no relation to the type of camera or lens used.

6: They are about the same. Maybe Frehner's photo shows a slightly larger wall, but they are very similar. Perhaps, Frehner was just a little further to the left of Meier's position.

7: They are very similar but not the same. Frehner made a grand effort, but his position just slightly shifted from Meier's. Maybe the garden bench prevented him from being a bit more to his right, closer to Meier's actual location. Point A does not align with the center of the window. Point D is to the right of point E, and so not in alignment. You noticed point E is at a higher elevation, almost at the same elevation as point D, but Meier's photo shows point E as lower. Since the elevation of point A on the window is about the same (albeit not centered) we think the elevation of Frehner's camera was about the same as

Meier's. The higher position of point E may have resulted from a new concrete or stone floor put in place post-1980, or possibly a slight difference in height between the photographers and their cameras. The ground differs in both pictures. Meier's photos do not show a concrete floor. The new floor may have elevated the ground around 10 centimeters. We have also assumed the little wall stones did not change their position, and upon comparing the photographic evidence, this seems to be the case.

8: Yes, the photographer and the carriage house are very clear. His camera is a bit below his eyes, and his location very close to the carriage house wall at one extreme of the garden bench.

9: He is very close to it. We estimate he is less than 1 meter from the carriage house wall.

10: No. **Rule c** in Experiment 1 demonstrates that changes in the sphere size do *not* affect proportional changes in reflected image areas. Here you see, at 1-meter distance, a huge carriage house wall, and with any size sphere it would be the same huge proportion. Skeptics have suggested taking a 55-centimeter diameter scale model WCUFO to the SSSC and photographing it to demonstrate whether a WCUFO model was or was not used by Meier. Skeptics are welcome to test a 55-centimeter model at the SSSC, but it would be a redundant exercise. All the spheres in Meier's photo show the same reflection, so testing any one sphere is sufficient, and the size of the sphere does not change the results. So, testing only one sphere of any size produces the same results as testing a scale model WCUFO in the SSSC.

11: If Frehner had walked about 5 meters away from the carriage house towards the right edge of Meier's house, the alignment of points D and E would not have changed much, possibly just a little in elevation. Point A would have changed its position more on the window since it would not have been in the direction of Frehner's movement. Point C would have changed more, moving very close to point B, even crossing to its right. The biggest difference would have been to the left end wall profile of Meier's home. Moving away from the carriage house would change these alignments. The plan views show this wall profile would have been far less visible, if at all, had the camera been at 5 meters from the carriage house rather than proximal to it. The only possible conclusion is that Meier was very close to the carriage house wall.

12: Our conclusions are:

- In most of Meier's photos alignments of the designated points do not change. Meier, therefore, was in the same position for every photo taken at the time.

- Meier's position is very close to Christian Frehner's in his experiment. We think Meier was around 60 centimeters to Frehner's right. (See Figure 26). We estimated the little stone wall shifted about 30 centimeters in Frehner's photos. As this wall is about halfway (19 m) between the camera and the right edge of Meier's house, the photographer's view shifts about twice as much as at the stone wall, or twice 30 centimeters. See Figure 27.

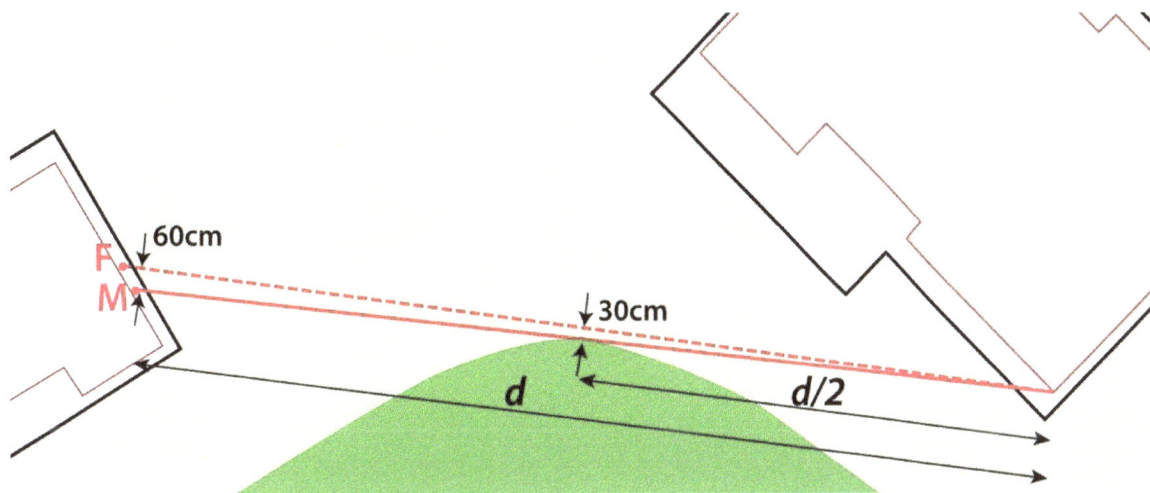

Figure 27- The shift between Frehner (**F**) and Meier's (**M**) view is about 60 cm or twice that of the stone wall.

Also, in this position, points E and D would be in the same alignment as in Meier's photos. Point A would be at the center of the window, in Meier´s photo too. Point C would be closer to point B, and the visible area of the left end wall elevation of Meier's house (red outlined area) would be a bit smaller, as Meier's photo shows.

- Definitively, Meier stood very close to the carriage house wall, and maybe around 60 centimeters towards the southeast of Frehner's test position.

- A sphere at just 1 meter away from the camera shows the photographer very clearly reflected, and a huge carriage house wall reflection. We calculate that the location of a 550 mm diameter scale model WCUFO would be just 1 meter from the camera, and the reflected image would be huge with the photographer and his tripod clearly visible. So we can know Meier did not use a 550 mm scale model. Also, in Figure 24 the sphere at 2 meters from the camera still shows the carriage house wall reflection far bigger than the images on Meier's photos (see Figure 28). So the spheres of this WCUFO were more than 2 meters from the camera. Experiment 4 calculates this distance far more accurately.

Also, based on **Rule b** experiment 1, we know the camera lens image must occur at the center of the sphere. See Figure 28, a zoomed image of a central WCUFO sphere for Photos #799, #800 and #808. The top row in this figure is the spheres without interpretation. The bottom row red circles define the spheres, and the blue dots are the sphere centers where we find the reflection of the camera and lens. The yellow profile is our interpretation of the carriage house wall with spherical deformation. Because an image in a mirror flips horizontally and a spherical mirror is no exception, when looking at Figure 26 we find Meier a bit to the left of the carriage house wall axis, indicated by the roof apex, and in Figure 28 the blue dots shift to the right of the rooftop.

You can check the other photos too, but consider that photos like #798, with the WCUFO on the right rather than towards the center of Meier's house, might, in their reflections, show the southeast wall of the carriage house in addition to the northeast wall. This more complicated reflection would make an accurate assessment more difficult.

Finally, in the detailed investigation of this WCUFO hovering just over the yard (Zahi *Analysis of the Wedding Cake UFO* 2013), we used an additional method to calculate Meier's location. So far we have considered alignments independent of the camera or lenses used. This additional method centers on the kind of camera Meier used. It is called the *camera formula* and concerns the focal length of the camera lens. In any photo, the size of the objects on the negative in the camera (or its sensor) is proportional to the scale of the actual object, and in the same proportion as the focal length of the camera to the distance from the object. We give simple, clarifying details on this in Experiment 7 (Chapter 9), "Distance estimation using the camera formula." In that method, to confirm the camera's location, you calculate the camera's distance to a known object on the front façade of Meier's home in the background. This camera formula method also confirms the camera was very close to the carriage house wall.

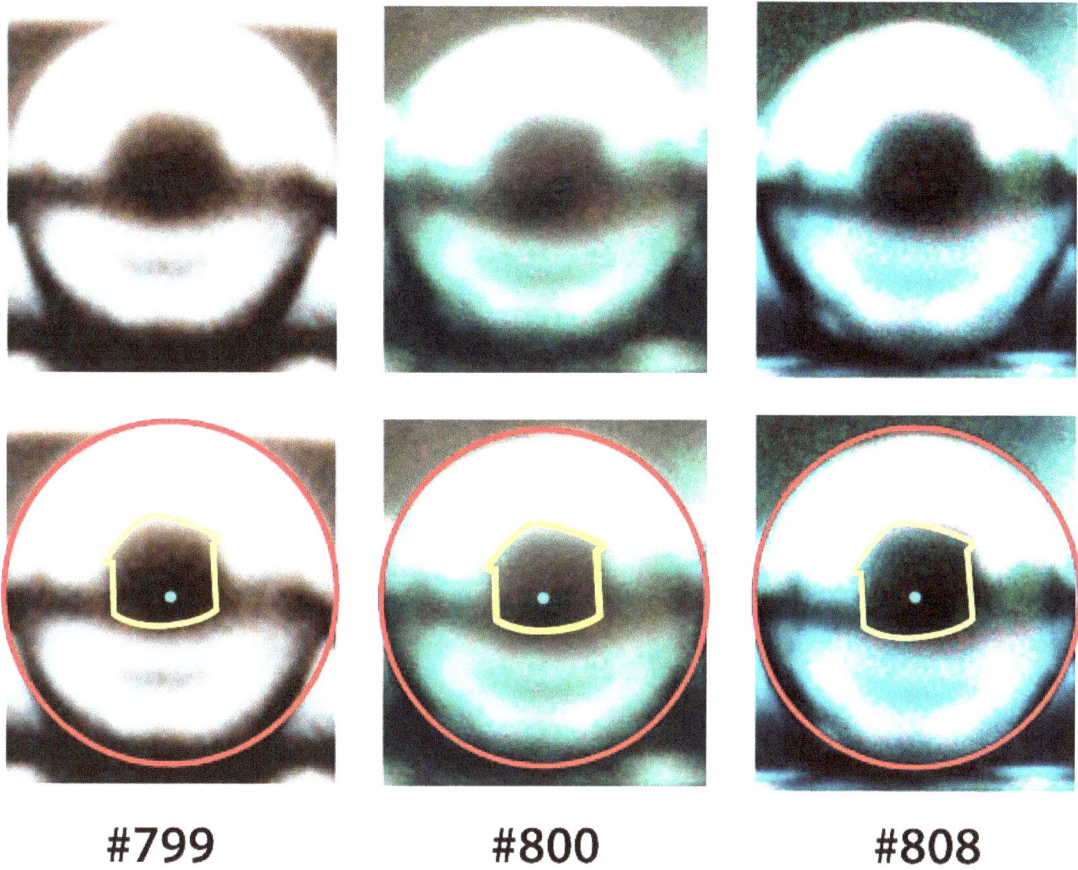

#799   **#800**   **#808**

Figure 28- Comparison of central sphere reflections on Meier's Photos #799, #800 and #808. The spheres' central blue dot shows the camera location. Sketched yellow profiles outline the reflected carriage house wall. Compare these reflection sizes and shapes with those at just 1 meter from the camera (Figure 25) or 2 meters (Figure 24).

Chapter 6

Experiment 4: Making and using a scale model of Meier's residential area

---

INTRODUCTION

This model puts you in Meier's shoes at the time and place and replicates taking his photos. It also enables experimentation with different WCUFO sizes and configurations.

MATERIALS

For this experiment, you need cardboard, glue, scissors, and a little reflecting sphere. Any reasonably sized sphere suitable for use in a constructed scale model of Meier's residential area is fine. Observations can be made with the naked eye, or a simple zoom camera, like a smartphone camera. The camera is useful because zooming the image enables clearer viewing of the sphere reflections and provides the option of recording your photographed observations.

METHOD

Making the model base

The first step is to use a scaled map which you can do in either of two ways.

    a. Print your plan view in large-format, longer than one meter.

    b. Use an available 3-meter or 1.5-meter map that we provide via a download (see Zip link below).

Know the scale of the plan view before printing, which makes it possible to measure distances knowing their length in the scale model. Conversely, measuring distances in your model, you then know their length in Meier's property. Figure 20 is an example of a simple map anyone can make and use. In this map, we included a yellow reference line marking out 10 meters (any suitable length is fine). Print your map (at a printing service with a plotter or something similar), then measure with a ruler how long the 10-meter yellow mark is (or your useful length). Suppose it measures 335 millimeters, then 335 mm to 10 m is your Scale Factor. In this case, at 1 m to 33.5 mm, your Scale Factor = 33.5 (millimeters in your model to 1 m in the real world).

Note this is your personal Scale Factor not an actual scale ratio, which in our case would work out to 33.5 in 1000 or 1 in 29.85, which rounds up to a scale of 1 in 30. Working on a computer makes it easier to use your Scale Factor on your model, but when talking to an audience, they can understand it more easily if given an actual scale number for the model. In our case, this is approximately 1 in 30.

The scale can also be in feet and inches. If measuring the size of the carriage house roof in feet, and making a yellow mark of 30 feet, the scale is the length of this yellow line in inches on your map, divided by 30 feet.

The scale enables distance calculations. To find a distance in your model from a distance in the real world, multiply it by the Scale Factor. For example, using your Scale Factor to measure 20 meters in the model, would be 20 m multiplied by 33.5, which equals 670 in millimeters in the model. The measurement can be confirmed using the scale ratio, 1 in 29.85. In our case, 20,000 mm divided by our scale 29.85 mm = 670 mm.

Conversely, to know how long something measured in the model is in the real world, divide it by the Scale Factor. For example, in your experiments on the scale model, to know how long your measured distance of 256 millimeters is in the real world, divide 256 mm by 33.5, which equals 7.64, in meters. Alternatively, again, just multiply the model measurement by your scale, here 29.85, which gives 256 mm x 29.85 = 7641 mm or 7.64 m. Checking with both methods confirms calculations, and a scientist always confirms findings. If available, two assistants or students could do one method each to confirm the measurements.

If you prefer not to use your map, choose the map that best fits your needs from the several scales available at this link:

www.rhalzahi.com/docs/Billy-yard-templates.zip

Figure 29 shows one of the maps available for downloading: the *big map* in color made from four sheets. You can select several templates from the downloaded file in two sizes or measurement systems (metric or imperial) and color or black and white. See the map templates available listed with their details in Table 1.

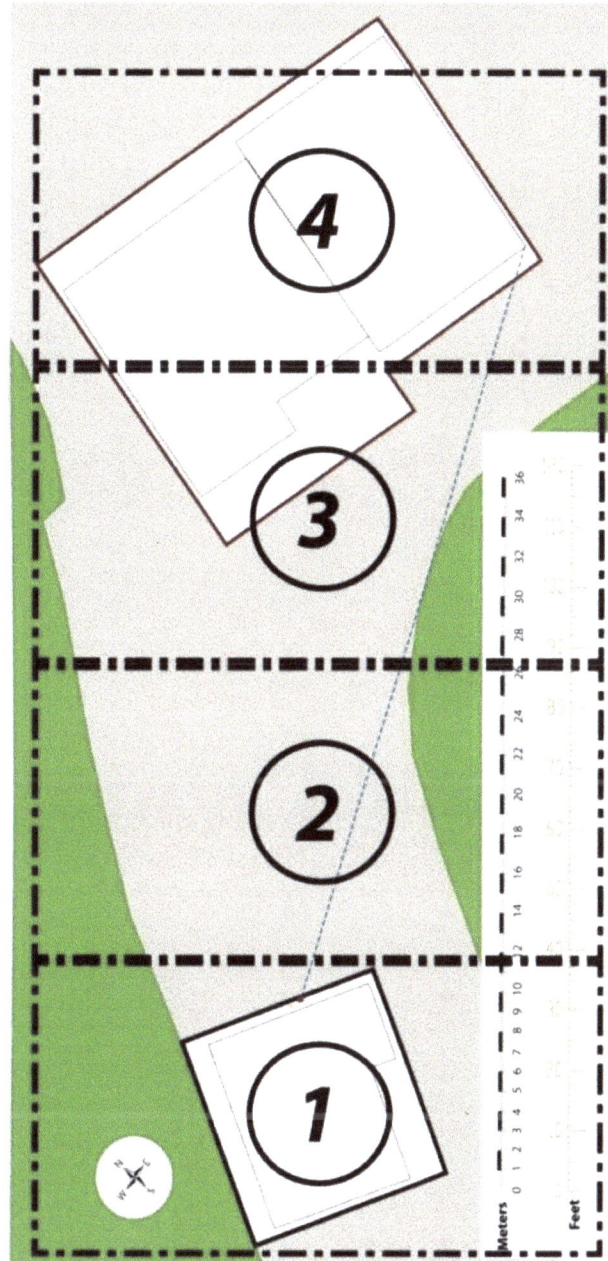

Figure 29- One example of the available templates. The *big map* made from four sheets printed and stuck together.

File name	Description
Map-Billy-Yard-small-metric-color.pdf	Color, metric system, small map in one single sheet, scale: 25 mm to 1 meter. (Or 1 in 40.)
Map-Billy-Yard-big-metric-color-part1.pdf Map-Billy-Yard-big-metric-color-part2.pdf Map-Billy-Yard-big-metric-color-part3.pdf Map-Billy-Yard-big-metric-color-part4.pdf	Color, metric system, big map in four sheets, scale: 50 mm to 1 meter. (Or 1 in 20.)
Map-Billy-Yard-small-metric-black.pdf	Black and white, metric system, small map in one single sheet, scale: 25 mm to 1 meter. (Or 1 in 40.)
Map-Billy-Yard-big-metric-black -part1.pdf Map-Billy-Yard-big-metric-black -part2.pdf Map-Billy-Yard-big-metric-black -part3.pdf Map-Billy-Yard-big-metric-black -part4.pdf	Black and white, metric system, big map in four sheets, scale: 50 mm to 1 meter. (Or 1 in 20.)
Map-Billy-Yard-small-english-color.pdf	Color, imperial system, small map on one single sheet, scale: 1 inch to 4 feet. (Or 1 in 48.)
Map-Billy-Yard-big-english-color-part1.pdf Map-Billy-Yard-big-english-color-part2.pdf Map-Billy-Yard-big-english-color-part3.pdf Map-Billy-Yard-big-english-color-part4.pdf	Color, imperial system, big map in four sheets, scale: 1 inch to 2 feet. (Or 1 in 24.)
Map-Billy-Yard-small-english-black.pdf	Black and white, imperial system, small map in one single sheet, scale: 1 inch to 4 feet. (Or 1 in 48.)
Map-Billy-Yard-big-english-black-part1.pdf Map-Billy-Yard-big-english-black-part2.pdf Map-Billy-Yard-big-english-black-part3.pdf Map-Billy-Yard-big-english-black-part4.pdf	Black and white, imperial system, big map in four sheets, scale: 1 inch to 2 feet. (Or 1 in 24.)

Table 1- Details of the different map templates available.

The *big metric map* measures 1.5 m by 3 m with the four sheets combined. The *big imperial map* measures 4 feet by 8 feet. The small maps of one single sheet, measure 75 cm by 150 cm, or 2 feet by 4 feet. So the *small map* can be placed on a table, but the *big map* requires the use of the floor. If using a camera, even a smartphone, ensure it can focus objects at a distance of 10 centimeters (4 inches) or more. If it can only focus beyond 20 cm (8 inches), it is better to use the *big map*.

Have your selected map printed at a printing service store after comparing prices for black and white, and color. If using the *big map*, stick the four sheets together.

Making the carriage house wall

Having printed your map, knowing the Scale Factor of your model, use it to calculate how big your carriage house wall is. Multiply every measurement on Figure 30 below by your Scale Factor. Draw it on a piece of cardboard and cut it out to look like Figure 30.

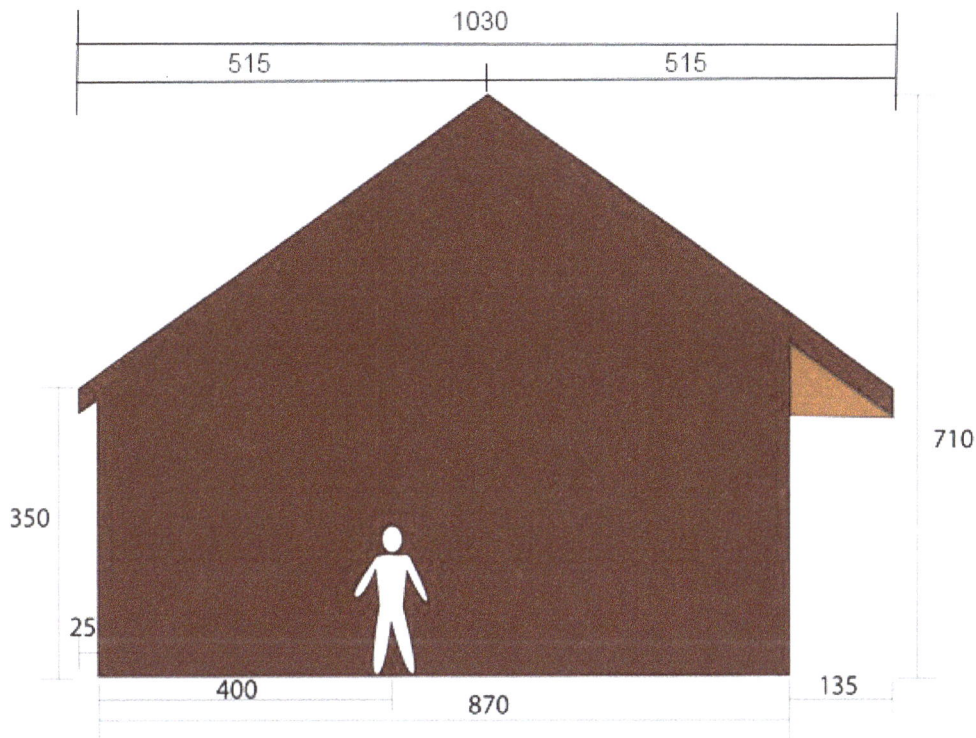

Figure 30- The carriage house northeast wall with dimensions in centimeters.

If, for example, your Scale Factor is 33.5, the base of the 8.7 m wall in real life measures 291 millimeters in the cardboard version (8.7 multiplied by 33.5). Confirmed by dividing your 8700 mm by the scale of 29.85 = 291 mm.

These carriage house dimensions measured on site, as shown in Experiment 2, are almost the same size we found in the Google Earth maps, so they are reliable.

There is a carriage house wall profile in the same Zip link as the available maps (see Table 2). Print it out, stick it on a piece of cardboard, and cut it out. Be sure to select the same size (*big* or *small*) and measurement system (metric or English/imperial) as the map you selected.

File name	Description
Carriage-House-small-metric-color.pdf	Color, metric system, small size, scale: 25 mm to 1 meter. (1 in 40.)
Carriage-House-big-metric-color.pdf	Color, metric system, big size, scale: 50 mm to 1 meter. (1 in 20.)
Carriage-House-small-metric-black.pdf	Black and white, metric system, small size, scale: 25 mm to 1 meter. (1 in 40.)
Carriage-House-big-metric-black.pdf	Black and white, metric system, big size, scale: 50 mm to 1 meter. (1 in 20.)
Carriage-House-small-english-color.pdf	Color, imperial system, small size, scale: 1 inch to 4 feet. (1 in 48.)
Carriage-House-big-english-color.pdf	Color, imperial system, big size, scale: 1 inch to 2 feet. (1 in 24.)
Carriage-House-small-english-black.pdf	Black and white, imperial system, small size, scale: 1 inch to 4 feet. (1 in 48.)
Carriage-House-big-english-black.pdf	Black and white, imperial system, big size, scale: 1 inch to 2 feet. (1 in 24.)

Table 2– Details of the different carriage house wall templates available.

Once the carriage house wall model is ready, make a circular hole in the face of the human figure, and attach supporting cardboard triangles to maintain the wall model vertical, as indicated in Figure 31.

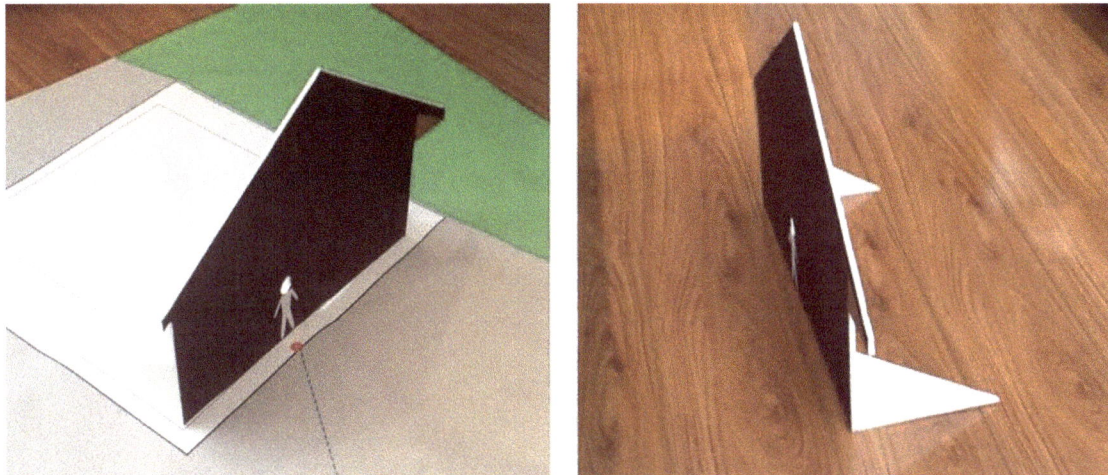

Figure 31- Model of the carriage house wall. Left: Positioned on the printed map with the hole in the wall where the face is. Right: Supporting triangles.

Assembly and alignments

Place your map on a table or the floor. Put the carriage house model on the north side of the corresponding map location (Figure 31, left).

Following are instructions for making the analysis of Photo #799, but the same applies to Photos #800 and #808. You can experiment with one or all of the pictures.

The next step is to locate three important points. The first is the projection of the left edge of the WCUFO on Meier's house in the background; the second is the projection of the center of the WCUFO on the house; and the third, the projection of the right edge of the WCUFO on the house. Two wooden sticks or a thick cord should suffice. To make a proper alignment the house must be observed in the background seen from the carriage house. You could temporarily put away the model of the carriage house, and from this position, place objects where you think each edge and the center must be. Referring to Photo #799 locate each point by looking at Meier's home from above, being careful not to make a mistake due to the perspective factor. It is safer to do this alignment from the carriage house location.

Figure 32 illustrates the process. Here we use two poles pointing at the left and right edges of the Photo #799 WCUFO. Once you are sure of these alignments, draw two lines converging to the photographer's location close to the carriage house.

In Photo #799, the left edge points a bit beyond the far left side of Meier's home roof, and the right edge goes to its front segment, to the right, as indicated in the figure below. Take your time to do these alignments checking they are right. If it helps, an assistant could place an object (a candlestick here) in the correct positions while you observe from the carriage house model.

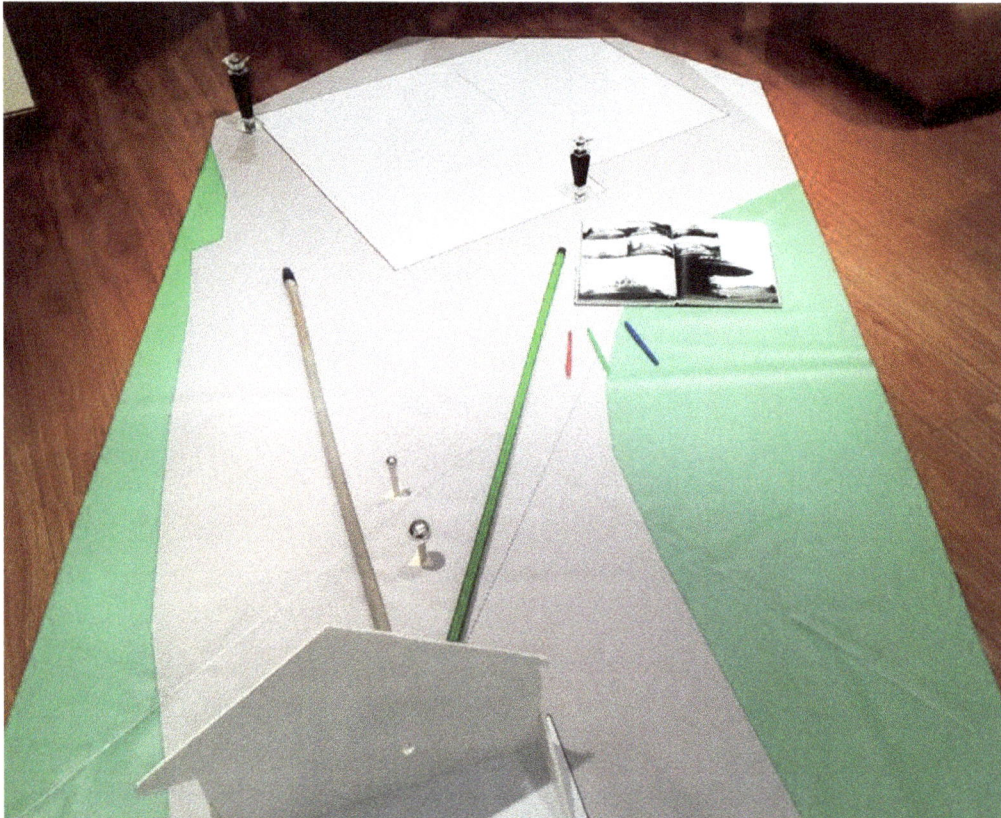

Figure 32- Photo #799 WCUFO: Two pole alignment of its left and right edges viewed from the carriage house. Two thick cords could also suffice.

Do the same for the central axis of the WCUFO. The central axis line from the observer must be exactly between the two poles accurately bisecting their lines pointing towards the left and right edges of the WCUFO.

Put a small reflecting sphere on top of any support to elevate it to the photographer's eye level on the carriage house wall.

The resulting lines appear in red In Figure 33. On the red dashed bisector, mark every half meter (or every foot) from the photographer towards Meier's home. The distances are real lengths in Meier's yard at scale. So, if using one of the available templates, use the scale on the side of the map. Alternatively, if using your map, calculate the distance with your Scale Factor (or your scale ratio). The marks range from 4 meters to 7 meters (or 13 feet to 23 feet).

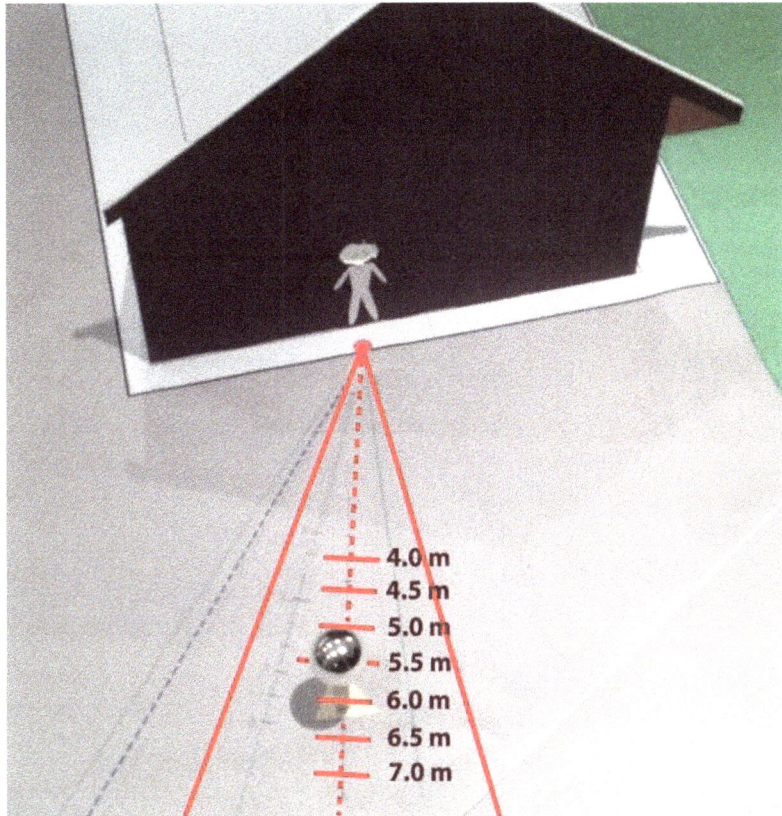

Figure 33- Projected lines from the photographer to the left and right edges of the WCUFO in Photo #799. The dashed bisector line goes to the center of the WCUFO. Mark every 50 cm, (or 1 foot) from the photographer.

Observing the sphere reflections

Put your camera behind the cardboard carriage house wall, with the lens at the hole in the face of the figure. If light, fix it with masking tape. Just ensure it points to the red bisecting axis illustrated in Figure 33. Alternatively, mount the camera on some books to elevate it up to the hole.

Do not worry about the smartphone camera being behind or in front of the carriage house wall. We know from sphere reflection **Rule a (ii)** that it does not matter where the camera is, the size of the reflected carriage house wall on the test sphere is the same.

Figure 34- Sphere photos taken from the model of the carriage house wall. Left: Taken at 1 meter from the photographer reveals a much larger reflection than in Meier's photos. Right: Taken at around 6 meters away. The red rectangle inset shows its sphere with a reflected carriage house image very similar in size to those in Meier's photos (Figure 27).

Take notes for each photo taken. Place your sphere on the first mark, at 4 meters in Figure 33, and take a photo. Then move the sphere to the next mark and take another photo. Do the same for all marks. You could take a picture of the sphere at one meter (3 feet) from the photographer, the distance a small model would be. You have now made photos similar to those in Figure 34.

Compare how similar the reflections are of the carriage house at just 1 meter from the camera and the test made by Christian Frehner at 1 meter (Figure 34 left and Figure 25). Also, note the similarity in size between the carriage house reflection in a sphere at 6 meters away (Figure 34 right) and its size in the spheres in Meier's photos below (Figure 35).

Now compare the size of the carriage house reflection in each of your photos with its size in Meier's photos. If you find the sphere at 6 meters shows the same size carriage house reflection as in Meier's photos, it means the WCUFO sphere was 6 meters away from him.

The following pages now show how to calculate both the reflection and the carriage house sizes and proportions to a good degree of accuracy.

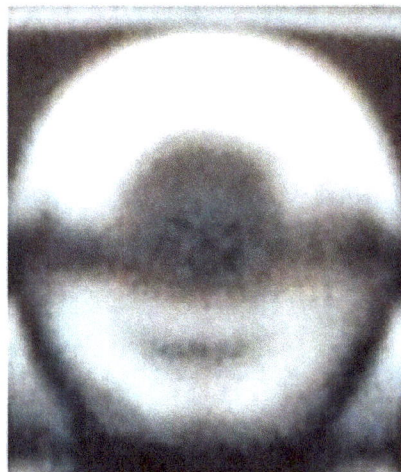 #799

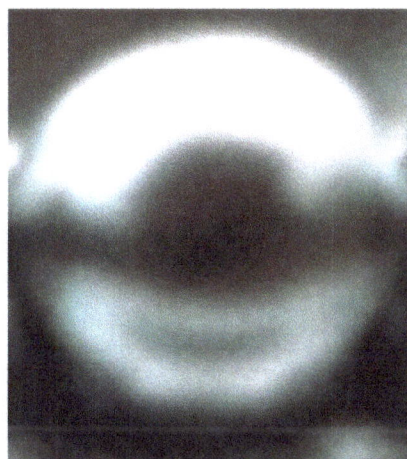 #800

 #808

Figure 35- Central sphere details on Photos #799, #800 and #808.

Simple calculations

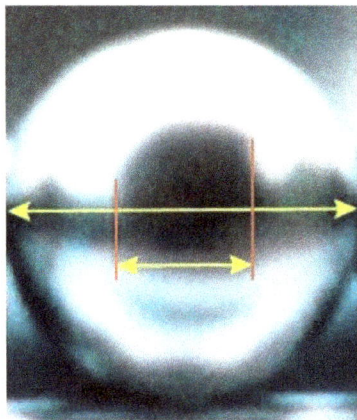

Figure 36- **Rch** calculation.

Figure 35 has zoomed images of the spheres at the center of Meier's Photos #799, #800 and #808. Photo #808 is the clearest one.

With a rule measure the width of the carriage house roof on this page in Figure 36. The measurement must go from the left roof eave to the right roof eave (bottom yellow dimension line). Measuring the wall width minus the building's eaves is hard since there were a few plants on each side of the carriage house, making it more reliable to base calculations on the roof width.

Now measure the full sphere diameter (top yellow dimension line) in Figure 36. Divide the roof width by the sphere diameter. This division we call **Rch** (carriage house roof ratio). For us, the roof width was 22 mm, and the sphere 60 mm. So we have:

$$Rch = \frac{22 \text{ mm}}{60 \text{ mm}} = 0.367 \text{ or } 36.7\%$$

Corresponding images of Figure 35 exist in various books. Use any of them or Figure 35 to calculate **Rch** for each photo. (See also Annex D.)

The above is just our example. Find *your* measurements, accurately measuring the sphere diameter from the picture under analysis.

Now, in a table like Table 3 below (or use this one), at the top write the date, the **Rch** value obtained from your measurement in Figure 35 and check the number of each photo you use. Upload your camera photos to a computer and zoom the sphere images on the monitor. Write at what distances (feet or meters) you take the photos you analyze, and using a ruler measure the Roof Width (eave to eave) and Sphere diameter in each picture. Then divide both values for each of your photos to obtain new **Rch** values. Record all these new **Rch** values under **R**.

Photo ☐ 799 ☐ 800 ☐ 808
Date:
Rch:

Distance from camera		Photo #	Roof Width	Sphere diam	R
Meters	Feet				
4.0	13				
4.5	14				
5.0	15				
5.5	16				
6.0	17				
6.5	18				
7.0	19				
	20				
	21				
	22				
	23				

Table 3- An example of a table used to record your results.

Find the **R**-value on your table closest to the **Rch** value. For example, if at 6 meters, the **R**-value is very close to the **Rch** value, then in this photo the WCUFO was situated where the nearest sphere to the camera was 6 meters away. If two **R**-values are very close to the **Rch** value, interpolate or take their average. If for example, these values were 5.5 meters and 6 meters; the distance would be the average of 5.75 meters. We call this distance **Ds**, the distance to the nearest sphere.

On a sheet of paper draw a small circle representing this nearest sphere at the **Ds** you have found (blue circle Figure 37). With it drawn, estimate as accurately as possible its WCUFO size and draw that (black circle Figure 37). Draw red lines bounding the WCUFO on its left and right sides. You now have a drawn diagram of WCUFO geometry on your map looking like Figure 37.

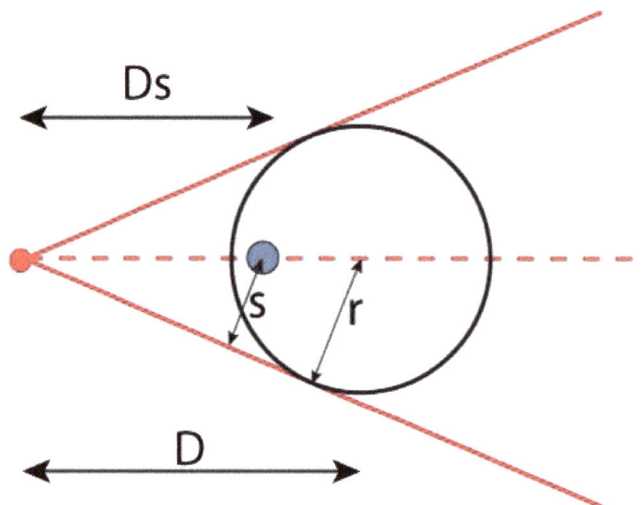

Figure 37- Diagram of WCUFO geometry: The blue dot is the nearest sphere; the red dot the photographer, Meier; and the red lines bound each side of the WCUFO, the black circle.

The drawing gives a good estimation of the WCUFO size and avoids math calculation for those weak at math. See Experiment 6 for a simple method of making several WCUFO circular paper models to find its real size.

A more complex calculation

Another way to calculate the WCUFO radius, half its diameter, is using a ratio formula. In Figure 37 **r** is the WCUFO radius. The line segment **s** is the shortest distance from the center of your previously drawn nearest sphere (the blue one) to one of the red bounding lines. Since the axis (red dashed line) bisects the red lines, you can measure **s** to either of them because they give the same value. These values may not be the same if your drawn red dashed line does not exactly bisect the lines. In that case, measure the distance from the sphere to each bounding line and take the average. The value of **s** must be in real courtyard units, so check the map scale to calculate the distance. **Ds** is the distance you found the sphere must be from the carriage house wall where Meier took the photos, and **D** is the distance from the camera to the center of the WCUFO.

In Figure 37 there are two similar triangles, so we have the proportion:

$$\frac{Ds}{s} = \frac{D}{r} \qquad (1)$$

Moreover, we know the sphere is located at 0.61 times the radius **r** from the center of the WCUFO (Annex A). So we may say:

$$D - Ds = 0.61\,r$$

$$D = 0.61\,r + Ds$$

Substituting in (1):

$$\frac{Ds}{s} = \frac{0.61\,r + Ds}{r}$$

$$r\,Ds = 0.61\,r\,s + s\,Ds$$

$$r\,(Ds - 0.61\,s) = s\,Ds$$

So:

Radius of WCUFO

$$r = \frac{s\,Ds}{Ds - 0.61\,s}$$

Calculating the WCUFO radius and diameter

If your **Ds** is 6.5 meters, and the distances measured from the sphere to the bounding lines (**s**) are 1.5 m and 1.3 m. The average value of **s** = 1.4 m

Substituting these values into the formula gives:

$$r = \frac{1.4 \text{ x } 6.5}{6.5 - 0.61 \text{ x } 1.4}$$

$$r = \frac{9.1}{6.5 - 0.854}$$

$$r = \frac{9.1}{5.64}$$

$$r = \mathbf{1.61}$$

Therefore, since the radius is half the diameter:

The WCUFO diameter is = 3.2 m.

To check these and other calculations more precisely see the book *They are Here* (Zahi and Lock). To use less math, drawing the WCUFO from the location of the sphere as illustrated in Figure 37 is fine. Alternatively, see Experiment 6 for yet another method.

As noted, any camera location between the carriage house and the sphere is acceptable. See **Rule a (ii)** in Chapter 3. The size of the reflected carriage house image remains the same no matter where the camera or Meier locate. If Meier used a small scale model at around 5 meters from the carriage house wall instead of very close to it, the carriage house sphere reflections would be the same size as in his photos. Does this mean Meier could have used a small model at 5 meters from the wall? No. We have already established in Experiment 3 that Meier had to be very close to the carriage house. We know he was not 5 meters away from it (for additional details see Zahi and Lock *They are Here* "Skeptics' Comments, Questions and Answers").

Finally, we have referred to Photos #799, #800, and #808, but what about the other WCUFO courtyard photos in Figure 5? In some of these other photos, Meier's house is not visible making it hard to define the bounding lines due to lack of background image reference.

Photo #803 could be analyzed, but the left side is not entirely visible. So we would recommend first defining the bounding line on the right, then the central bisecting line, and finally projecting the left bounding line at the same distance and angle from the bisector as on the right.

Photo #798 shifts to the right, making not only the main carriage house wall visible but also a side wall and more roofing. In that case, a model of the entire carriage house, not just one wall, must be used.

Chapter 7

Experiment 5: Making and using big scale models of Meier's site & camera

---

To investigate the WCUFO some people may think it essential to visit Meier's home and test various sizes of WCUFO models there. Not everybody, however, can travel to Switzerland and perform such research on his property. So an alternative way is to construct a simple full-size replica of Meier's courtyard and carriage house. We could do this with the maps from Experiment 4, but its resources may be impractical or too expensive for some people.

Scientists wanting to investigate how the Titanic sank would not construct a replica of the ship and send it across the North Atlantic to collide with an iceberg. Scientists use different or alternative methods. We have already employed an indirect method by using a small model of Meier's courtyard in Experiment 4. Now we take another approach by making a full-size model: a field model. We also make a very simple large scale cardboard model of Meier's camera.

This real field model helps you experience Meier's physical location and the geometry of his camera. The experiment requires an open space outdoors, and some basic resources, yet it is simple and fun for a small team to conduct.

INTRODUCTION

In this experiment, you learn how to determine the WCUFO size and location for Photos #799, #800 and #808, and how to make a full-size test model (1 to 1 scale) by simulating the full-size carriage house roof. With the help of three volunteers, you confirm the reflection size in a Test Sphere.

You also construct a simple large-scale cardboard model of Meier's camera, not to shoot photos, but to establish the WCUFO size and Meier's distance from the craft when taking these photographs. It simulates the geometry not the optics of Meier's camera. There is no need to buy any film or photographs, and it too can be fun to use.

MATERIALS

For this experiment, prepare the following materials:

- **A large reflecting Test Sphere** to take on site. The bigger the sphere, the better the image. The size is not critical because any size produces the same results (sphere reflection **Rule c**). The sphere requires a setup that we explain in "Test sphere preparation" below.

- **Some white tape** to mark your sphere, and for backup.

- **A model of Meier's camera**, with the same geometrical proportion as Meier's camera used for taking the WCUFO photos. For this camera model, you require: a piece of rigid cardboard, dark or painted black on one side; some 200 mm by 140 mm transparent slides from a local store or cut by yourself from sheets; a flat screen computer; and a felt tip pen. We explain how to create this cardboard camera model below.

- **At least three volunteer assistants**. Ask them to dress in light colored clothes. Alternatively, take two white blankets or sheets to drape over them, so they contrast well with the local environment and become clearly visible in the Test Sphere reflections.

- **At least six stakes or little sticks** to drive into the grass, making key field markers on site.

- **A tape measure** or a somewhat flexible meter ruler for measuring field distances and distances around your Test Sphere perimeter.

- Around **20 meters of cord**.

- Any **camera with zoom capability**, and if it lacks zoom capabilities, a pair of binoculars too.

- **A notebook** in which to make annotations.

- **A copy of Table 4** to take on the field.

Preparations

Test Sphere preparation

Find a large reflecting sphere, between 20 cm to 40 cm in diameter (8 to 16 inches). You need to stick white vertical strips of masking tape, or something similar, on this Test Sphere, as in Figure 38. A permanent marker is problematic if mistakes occur. These white strips represent the left and right sides of the carriage house roof reflection.

Stick these strips of tape on your Test Sphere in advance. In our experiment, we had the strips for each test photo very close to each other on the Test Sphere, but it might be better to separate them around the sphere with two strips for each photo and to write the photo numbers at the top of the strips. Using too many strips close together may cause confusion.

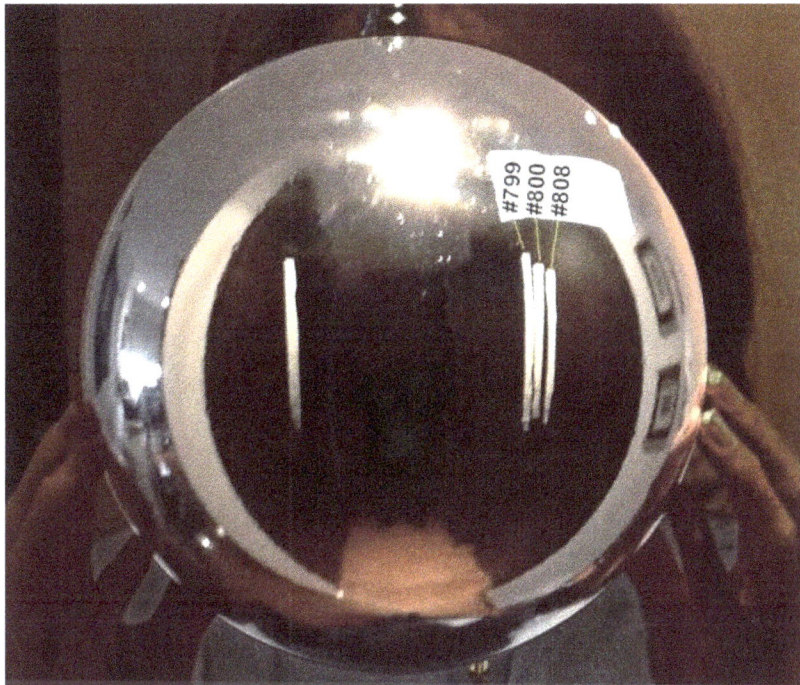

Figure 38- Test Sphere with white masking tape strips for each photo analysis. The left strip is common to all three photos while the right-hand strips each denote one of the three photos. The wider the gap between the left and right white strips the closer the WCUFO was to the camera.

Math reveals the distance between each pair of white strips on your Test Sphere. If you like math, you can confirm the formula we obtained or look it up in Annex C. Alternatively, avoiding math you can rely on the following simple function below to calculate the distance or ask a mathematician for confirmation.

Let **L** be the distance between the strips for each pair of left and right white strips on the Test Sphere. **L**'s length depends on the circumference or perimeter **P** of your sphere. To make all the surface measurements on the Test Sphere, you need a flexible meter ruler or tape measure. For the perimeter **P**, measure the largest circumference around the sphere, or knowing its radius or diameter calculate **P** = Pi (3.1416) x 2r.

To calculate the value of **L** we use this formula:

$$L = \frac{P}{180°} \, Sin^{-1}(Rch)$$

where:

L   is the distance between your strips (each pair of white strips) measured on the sphere surface.

P   is the perimeter (largest circumference) of the sphere.

Rch   is the reflected horizontal roof width divided by the sphere diameter. Here, the decimal value not the percentage.

Sin⁻¹ is the Arc Sine or inverse function of Sine. The resulting value must be in degrees.

L units in this formula are the same units used for your perimeter **P**.

For example, for a Test Sphere perimeter of 625 mm. For Photo #799, we calculated a **Rch** value of 0.336. So we have:

$$L = \frac{625\ mm}{180°} \, Sin^{-1}(0.336)$$

$$L = \frac{625\ mm}{180°} \, 19.63°$$

$$\mathbf{L = 68\ mm}$$

So, in this example, the two white strips are separated 68 mm on the sphere (Figure 38).

Use white strips for all three photos. First stick a strip of whitish masking tape on the left, then with the flexible ruler or tape measure, locate the other three strips at their respective distances **L** from the left strip by measuring **L** on the Test Sphere surface. Notice that **L** for #799 is smaller than for #800 and #808. Since the WCUFO was coming closer to Meier, the value of **L** gradually increases.

In the field, during the experiment, two volunteers, **V1** and **V2** (Figure 43) make a *perfect match* to these strips (see Figure 45).

Take extra masking tape to the field for your test, just in case the tape strips fall off the Test Sphere during transportation or testing.

Making a model of Meier's camera

If we know the WCUFO in Photo #800 has an angle of view of 28° what does this mean? Similarly, if we say, Meier's camera lens has a focal length of 55 mm, what does this number imply? They may have technical meanings for people skilled in photography or geometry, but for others, they may mean nothing. By making a simple camera model from cardboard, however, and using translucent templates representing each photo, these concepts can take on practical meaning for almost anyone.

How does this model of Meier's camera work? There are two ratios in camera geometry that we need to incorporate into any camera model we make. The first one is the relationship between the width and height of the film negative, which is 1.5 to 1. Roll film negative like Meier used measures 35 mm by 23 mm, which is a ratio of virtually 1.5 to 1. So any camera model must have an open end of this ratio.

Figure 39 shows a simple cardboard model camera. It is in a pyramidal shape with a square open extreme of 10 mm by 10 mm at the top, and an open base rectangle of 180 mm by 120 mm, the same ratio as a 35 mm photo: 1.5 to 1. The distance from the eye of the observer to the rectangular opening is the same ratio as Meier's camera focal length (55 mm) to the size of the 35 mm roll film negative.

Such a cardboard model camera can be made using the template illustrated in Figure 40. The template sheet size is 800 mm by 710 mm. As the figure shows, our camera model template has an open rectangle of 180 mm by 120 mm, the essential ratio of 1.5 to 1.

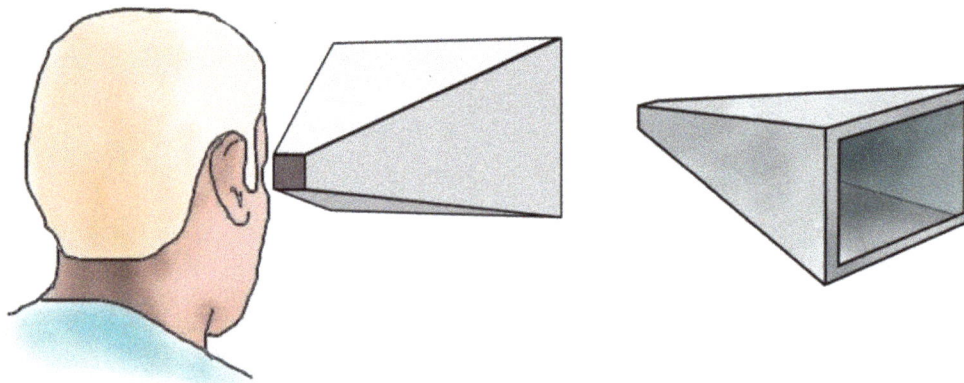

Figure 39- A simple cardboard model of Meier's camera with the geometric proportions of a 35 mm roll film camera with a focal length of 55 mm.

So, how far away from this rectangle must an observer be to have the same angle of vision as Meier's camera? To see in the same ratio as a wide angle lens camera in the camera model, the observer would need to be very close to the 180 mm by 120 mm rectangle. Then, with an eye very close to the open rectangle a wide view would be seen. Conversely, for a camera with a telephoto lens, which produces a very narrow-angle view, the observer would have to be far away from the 180 mm by 120 mm rectangle. So being closer or farther away from the open rectangle changes the angle of vision and how much of the scene you see.

Meier used a 55 mm focal length camera. So we need to know how far away the observer's eye (small square top of the pyramid) must be from this 180 mm by 120 mm opening to have the same angle of vision. For this, we need the other ratio: the ratio between the 55 mm focal length of Meier's camera lens and the 35 mm roll film width. In other words, a ratio of 55 mm to 35 mm, or 1.57 to 1. So if our camera model has an open width of 180 mm (template in Figure 40), the eye must be located at 1.57 times this value from the opening to give the same angle of view as Meier's camera. Multiplying 1.57 by 180 mm gives 283 mm. The template we designed has this ratio in a pyramidal shape. We allow an eye relief distance of 10 mm. So the designed template gives a perpendicular distance of 283 mm from the 180 mm by 120 mm surface to the observer's eye. Note the template 281 mm is the diagonal length of the pyramid not the perpendicular distance to the opening which is 273 mm.

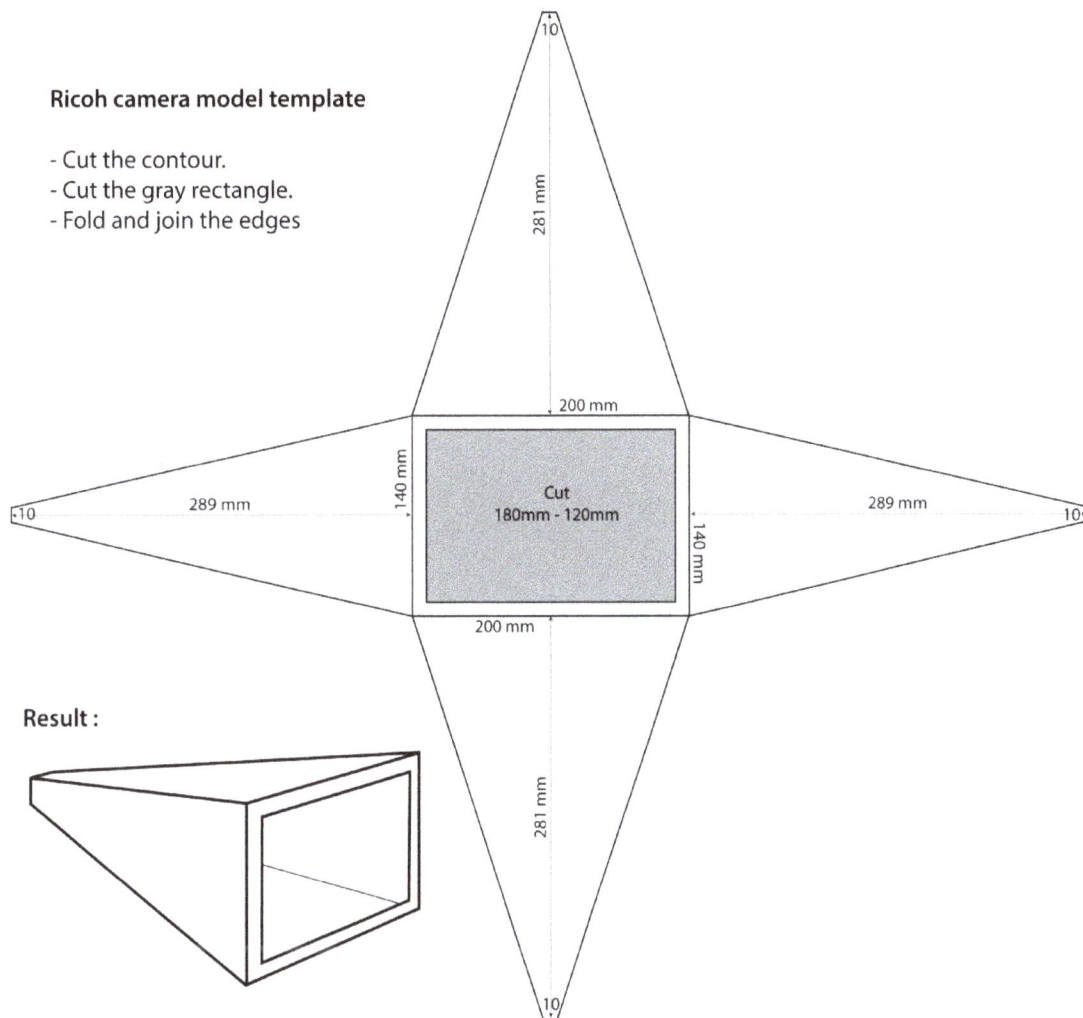

Ricoh camera model template

- Cut the contour.
- Cut the gray rectangle.
- Fold and join the edges

10

281 mm

200 mm

140 mm

-10   289 mm   140 mm   Cut
180mm - 120mm   140 mm   289 mm   10·

200 mm

Result :

281 mm

10

Figure 40- Template to construct the cardboard camera model.

You can download the template here:

http://www.rhalzahi.com/docs/Camera-template.pdf

Now looking through the open rectangle of 180 mm by 120 mm via the little square hole in the top of the cardboard pyramid, you see the same restricted world view (the same angle of vision) as Meier's camera. This same exterior view is important because we can draw transparent slides with WCUFO images and put them on the camera model opening, thus enabling the same angle of vision projection from your camera model as the WCUFO photos. This angle of camera view together with the following sphere test experiment tell us how far away the WCUFO was, and this gives us a sense of just how big the object was.

Print out this template, or use the above figure to construct yours. A dark cardboard would be best or paint the inside black before assembly. Cut it out, extract the inner rectangle, fold it and join the sides with transparent tape or glue. You use this camera model by looking through the small square top opening at a WCUFO scaled drawing on a transparent slide affixed to the large end, as illustrated in Figures 42 and 46.

Find some transparent slides 200 mm by 140 mm, or make your own from some transparent sheet.

Figure 41- Outline tracing of a WCUFO on a transparent slide over the PC image, to use with the Meier cardboard camera model.

Upload the photo for testing onto your PC screen. Zoom the size until it is 180 mm by 120 mm, the size of the rectangular opening in your Meier cardboard camera model. Center the transparent slide, fix it with transparent tape to the computer screen, and with a permanent marker draw, or trace, the WCUFO outline on the slide (Figure 41). Draw an outline tracing for other WCUFO photos on these slides to use with your camera model.

During the tests, you fix these slides to the end of the camera model at the big rectangular opening and then look through the small square top opening as indicated in Figures 42 and 39.

You can download the photos included in the following Zip files to make the required templates:

http://www.rhalzahi.com/docs/Pictures-Tree.zip

http://www.rhalzahi.com/docs/Pictures-Yard.zip

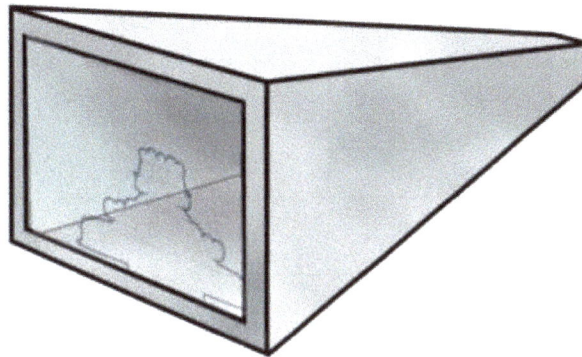

Figure 42- Transparent slide with the WCUFO outline on the big opening of the cardboard camera model.

We used this camera model for our outdoor tests of Meier's WCUFO courtyard photograph and his photo of a WCUFO proximal to a tree (Figures 46 and 47).

Preparation summary

You are now almost ready to go outdoors and experiment on site with friends or students, but first, in addition to the previously mentioned needed materials, finally prepare and check the following with everyone:

- Measure the **Rch** value on each photo to test. The **Rch** value is the ratio, or percentage, of the horizontal roof extension to the sphere diameter (Figure 36). You have these values from your calculations in Experiment 4. If not, measure them in the Figure 35 photos.

- Table 4 below is needed on the test site so make a copy. Fill in the **Rch** values you previously calculated from Figure 35. If you measured different values of **Rch**, use them instead of the ones below.

- Next look at the plan view of the outdoors experiment in Figure 43 and decide on distances **a** and **b** to locate the baselines. Distance **a** is from you, the photographer, to volunteer 2 (**V2**). Distance **b** is directly forward (at 90 degrees) from any point on **a**. Any distance for **a** and **b** are fine as long as they are at right angles to each other. For ease of operation, we made **b** 8.0 m in all tests.

- Because you observe the sphere reflections on the Test Sphere that could be some meters away, it is preferable to see a zoomed image with your camera or binoculars. Alternatively, you or another volunteer could move closer to the reflecting Test Sphere to check for the *perfect match* (Figure 45) when performing the experiment.

Photo	a	b	Rch
799	2.00 m	8.0 m	0.336
800	3.23 m	8.0 m	0.347
808	3.23 m	8.0 m	0.370

Table 4- Values of a and b for locating the baselines. Rch are our calculated values or use your own.

METHOD

Before going onto the field, sketch out, and go over your test runs indicated in Figure 43 to familiarize yourself with the process. Note that it is easier and more accurate to use the roof's horizontal width of 10.3 m reflected in the WCUFO spheres than the width of the carriage house wall.

Look at Figure 43. Two volunteers, **V1** and **V2**, stand 10.3 meters apart (33.8 feet). Each represents an extreme of the carriage house roof. You, representing Meier, stand on a straight line between the two, 4.25 m from **V2** at your right. We already know from Experiment 2 that Meier stood very close to the carriage house wall and also that the size of its reflected roof image on the Test Sphere does not change even if Meier was away from or behind the wall (sphere reflection **Rule a (ii)**).

Your third volunteer, **V3**, holds the Test Sphere in front of you while you stand on a line between **V1** and **V2** whose reflections representing the edges of the carriage house roof appear in the sphere image. The Test Sphere traverses back and forth along the baseline with **V3** until the affixed tape strips on the sphere match the positions of **V1** and **V2** (see Figure 45).

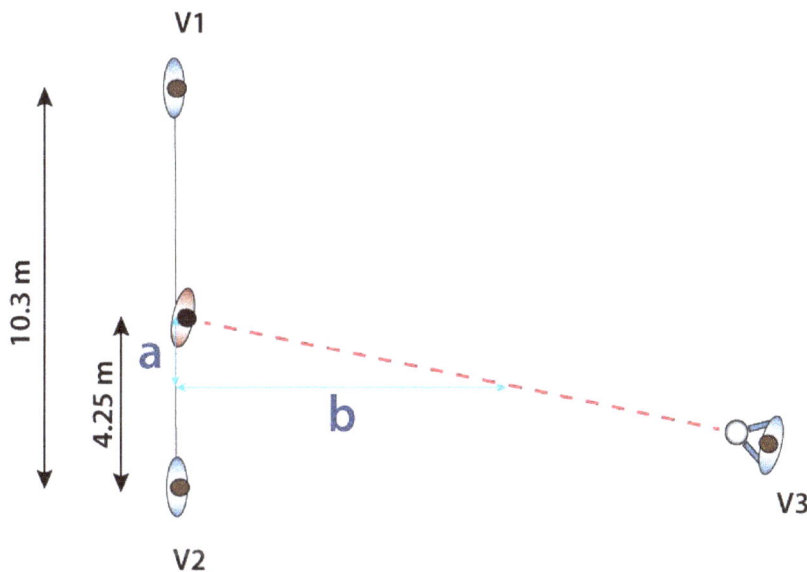

Figure 43-   Plan view of the outdoors experiment with red dashed baseline.

Now, test any or all of Photos #799, #800 and #808. The red dashed baseline in Figure 43 goes from you to the central and nearest WCUFO sphere in each photo, represented here by the Test Sphere held by volunteer 3 (**V3**). Since the photographs show the WCUFO was moving, this baseline direction varies from photo to photo. So there are three different baselines, one for each of the three photos. The WCUFO orientation in Photos #800 and #808, however, is about the same; the WCUFO central axis points to the same part of Meier's home. So in practice, we have just two baselines, one for #799, and the other for #800 and #808. Then, to verify how the WCUFO was moving above Meier's parking lot, you mark on the ground with a stick or stake where it was in each photo. These baselines later guide you to finding not only the location but also the size of the object.

After having determined the location of the WCUFO's nearest sphere, in the second part of this experiment we use the cardboard model of Meier's camera to observe from very near the simulated carriage house wall to determine how big this WCUFO was. We observe from the same location Meier took his photos. It is important to observe from here because the size of the WCUFO depends on the position of Meier's camera not the size of the carriage house reflection. You already proved in sphere reflection **Rule a (ii)** that that is the same wherever the observer is along the baseline between the carriage house and the Test Sphere. Doing this, we can obtain a reliable estimate of the WCUFO size and how it moved within the front yard.

Part One: Establishing the Test Sphere location on site

Enlist the help of at least three volunteers. When first arriving at the field or park find an open space for your test site. A flat terrain is preferable but not essential. In practical terms, the height of the sphere does not change the results. Referring to your sketches or copy of Figure 43 proceed as follows:

- Use two stakes to locate volunteers 1 and 2 (**V1** and **V2**) 10.3 meters apart (33 feet 10 inches). Check that, if possible, a dark background is behind you.

- Use another stake to mark your location between **V1** and **V2**.

- To locate each baseline on the ground:

  Use the two distances **a** and **b** that you decided on and entered into Table 4 for each baseline. Measure your distance **a,** towards **V2,** from the stake placed where you stand with your camera. Then take your distance **b** from Table 4, measure it in front of you and drive in a stake. With a cord connect this new stake to the stake where you are observing. Repeat the same procedure for the other baseline. Finally, you have two ropes or cords, one for each baseline, going away from the stake where you stand.

- Ask volunteer 3 (**V3**) to hold the prepared reflecting Test Sphere at around two meters from you.

  With **V1** and **V2** in position, have **V3** gradually move back along your baseline cord on the ground. **V3** rotates the Test Sphere to the left or right to assist a clear view of the four strips and both **V1** and **V2** if necessary. With binoculars or the camera zoom, check the alignments of **V1** and **V2** with the sphere strips. If the Test Sphere is too close to you, reflections of **V1** and **V2** are wider than the corresponding sphere strips. If it is too far away, they are narrower, and the volunteers' images are closer together than the strips. Figure 44 illustrates the procedure. When the exact distance separating the two volunteers matches the strips perfectly on the sphere, you have the same distance as the WCUFO sphere to the carriage house wall. Ask **V3** to drive a stake in the ground vertically below the center of the sphere. See our test for Photo #799 in Figure 44.

Figure 44- Test Sphere reflections. The WCUFO position in Photo #799 was analyzed using the far left white strip and first right white strip on the sphere. In this example, the Test Sphere must move back until the two volunteers align with the two white strips.

In Figure 44 the Test Sphere is too close, so **V3** must move back with the Test Sphere until **V1** and **V2** perfectly match the positions of the white strips. See also the simulation in Figure 45.

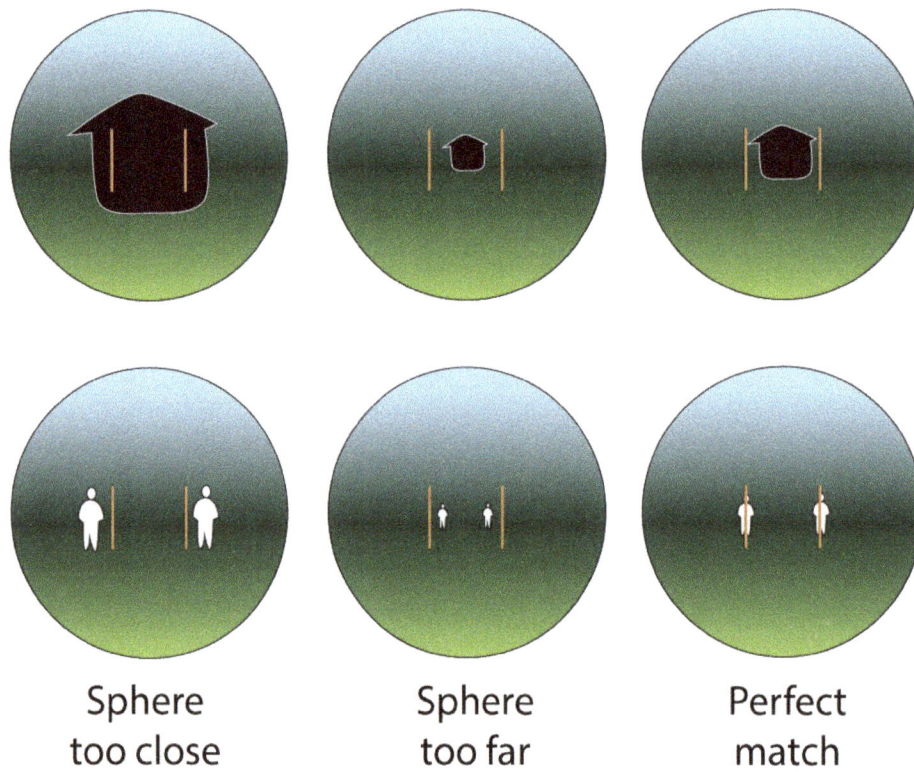

Figure 45- Simulated Test Sphere reflected images. Top row: Carriage house walls and replicas. Bottom row: Volunteer 3 with the Test Sphere traverses the baseline until the two volunteers' reflections fall perfectly on the sphere strips. The Test Sphere position then equates to that of the WCUFO center sphere.

It is safe to move closer to the Test Sphere, which makes it easier to check the reflections of **V1** and **V2** against the sphere strips. Sphere reflection ***Rule a (ii)*** tells us the results do not change however close or far we are to the sphere from the carriage house wall. In our test, we moved closer which explains the big photographer compared with the small volunteers in the Figure 44 reflection.

- With the Test Sphere positioned correctly and a stake in the ground beneath it, measure the distance between this stake and the stake marking your camera's position on the line between **V1** and **V2**. Note down this value. It is **Ds**, the distance to the nearest sphere.

Part Two: Establishing the WCUFO size

Now we use the cardboard camera model.

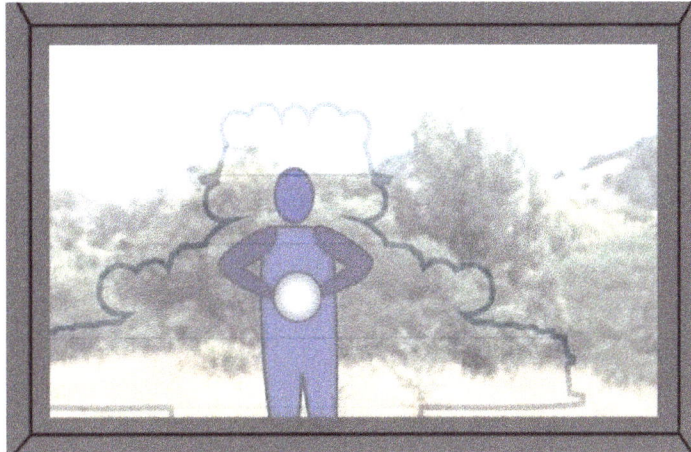

Figure 46- View from the cardboard camera model, with an affixed transparent slide of a WCUFO photo. Move the camera model until the Test Sphere aligns in place with the central WCUFO sphere.

- Put the transparent slide with the WCUFO outline from the photo you are testing on the front of the camera model. Look through the small square end (Figure 39) and move the camera model until the Test Sphere held by **V3** above the stake aligns in place, but not necessarily size, with the central WCUFO sphere. (See Figure 46.) Your camera must be in the observer´s first location, between **V1** and **V2**, at the carriage house wall.

- Keeping the camera model still, have **V1** and **V2** go up towards the WCUFO along the bounding lines to form a straight line with **V3** centered on this new line between them. (See Figure 52 for a plan view of the **V1** and **V2** positions where the bounding lines meet the WCUFO.) All three are still closer to the camera than the center of the WCUFO which is directly behind **V3**. Drive stakes into the ground at the new **V1** and **V2** positions. If one of the WCUFO edges is not visible (as in Photo #800, showing no left edge), one side only can suffice. With the **V1** and **V2** stakes in the ground aligned with the **V3** stake positioned at your Test Sphere, you can now estimate the WCUFO size using the **V1**, **V2** and **V3** locations.

- For more accurate results, use the formula already explained in Experiment 4 that gives the WCUFO radius. Just measure the distance **S** from the Test Sphere stick perpendicular to the line you use going from the camera to the right or left edge of the WCUFO (Figure 37). Use the previously calculated distance **Ds** from the photographer's location on the carriage house wall to the Test Sphere (Figure 43). Then use the formula arrived at from Figure 37, page 54:

$$r = \frac{s\ Ds}{Ds - 0.61\ s}$$

Where:

r   is the radius of the WCUFO base.

Ds   is the distance from the Test Sphere to the photographer close to the carriage house wall (Figure 37).

S   is the perpendicular distance from the Test Sphere marker to any of the WCUFO bounding lines as indicated in Figure 37.

- Multiply the WCUFO radius by 2 to obtain its diameter.

- Finally, repeat the same procedure for the other photos, using the corresponding baseline and strips on the Test Sphere. Since the separation of the strips on the Test Sphere is different for each photo, you obtain various locations for the WCUFO Test Sphere for Photos #799, #800 and #808. The stake for Photo #799 is the farthest away, and the stake for #808 is the closest. Stakes for both #800 and #808 fall on the same baseline.

- We know from sphere reflection **Rule c** that the size of the Test Sphere does not change the results, but a big sphere makes it much easier to see the reflections and gives greater accuracy. Note too that when checking alignments of **V1** and **V2** with the white strips on the Test Sphere image you (only you or an independent observer) can be as close as you want to the sphere and the result is the same.

Alternative use of the camera model

Finally, we perform another test with the camera model, this time on Photo #844 showing a WCUFO in front of a large but young Norway Spruce tree (Figure 47, top). Meier confirmed this tree height for the authors at 6 to 8 meters in a recent September 2016 telephone conversation with Frehner, which incidentally means the 15-meter height or *15 Meter hoch* in *Photo-Inventarium* on page 111 is incorrect. Analyzing this picture we found the WCUFO proximal to the 6 to 8 meter tall Norway Spruce. The WCUFO is so close that it projects a huge shadow over many of the tree's branches. In Experiment 9 you perform image processing of this picture and confirm this. From the picture, we also notice the WCUFO diameter is about half the tree height. So, being directly in front of the tree this WCUFO is approximately 3.5 meters in diameter.

Holding the cardboard camera model with the transparent slide of Photo #844 affixed to its big end, venture outdoors and find an equivalent 7-meter tall tree. Looking at it through the camera model move closer to, or farther away from it until its image is similar in size to the one in Photo #844. At this point, measure your distance from it, and so the WCUFO, to the camera. Our estimate is that Billy Meier was around 16 meters from the tree and WCUFO when he took this photo. See Figures 47 and 67.

Try experimenting with different WCUFO photos. Books, both old and new show dozens of pictures of this type of UFO. We have found the WCUFO comes in, at least, two different sizes, one around 3 to 3.5 meters, and the other around 7 meters in diameter. In your experiments, you may confirm this or perhaps make even more accurate assessments.

Figure 47- Cardboard camera model test of Photo #844. Top: Original Meier photo. Middle: Equivalent 7-meter tall tree. Bottom: Image through the camera model with an affixed traced transparent slide from Photo #844.

Chapter 8

Experiment 6: Calculating the WCUFO angle of view and size

INTRODUCTION

In Experiments 4 and 5 we calculated the distance from the camera to the nearest WCUFO sphere (**Ds**). We also used small and big scale models of Meier's site and calculated the size of the WCUFO. Moreover, we introduced the concept of the "angle of view" that a camera gives. In Experiment 4, Figure 32, the angle of view was the angle formed by the two wooden sticks pointing towards Meier's home.

The angle of view forms in the camera lens with the two bounding lines to each side of the WCUFO. If this angle is large, it means in general, that the WCUFO is close to the camera, while if the angle is very narrow and the WCUFO looks small in the picture it is far away from the camera. If a small object is very close to the camera, however, it can also give a large angle of view; so we need to know **Ds**, the distance to the WCUFO, to know whether the WCUFO is large or small. Because we found **Ds** in Experiments 4 and 5, we can use it together with the angle of view to estimate quite accurately the size of the WCUFO. It is easy to determine the angle of view because it is a function of the WCUFO size on the photo and the known camera characteristics such as its focal length (55 mm) and the size of the negative film (35 mm).

So, in this experiment, we expand on the angle of view by coming to understand its relation to the size of the WCUFO. Then, knowing **Ds** and the angle of view we fairly accurately estimate the WCUFO's real size. We also use the angle of view to determine distances for different sizes of WCUFO and to tabulate findings. You follow up by applying the process to other photos.

MATERIALS

For this experiment you require:

- A couple of big white sheets of paper 100 cm by 70 cm. Alternatively, use letter size paper, but results are more prone to errors with small sheets and taking measurements from small drawings.
- A long ruler to measure distances and draw lines.
- A pencil and colored markers.
- A protractor to measure angles.
- WCUFO photos from books or Annex D. If available, digital photos displayed on your flat screen computer are fine.

METHOD

The following method for calculating the angle of view is a visual method. After this, we look at an additional mathematical method.

Shown here is an example of measuring the angle of view for Photo #808.

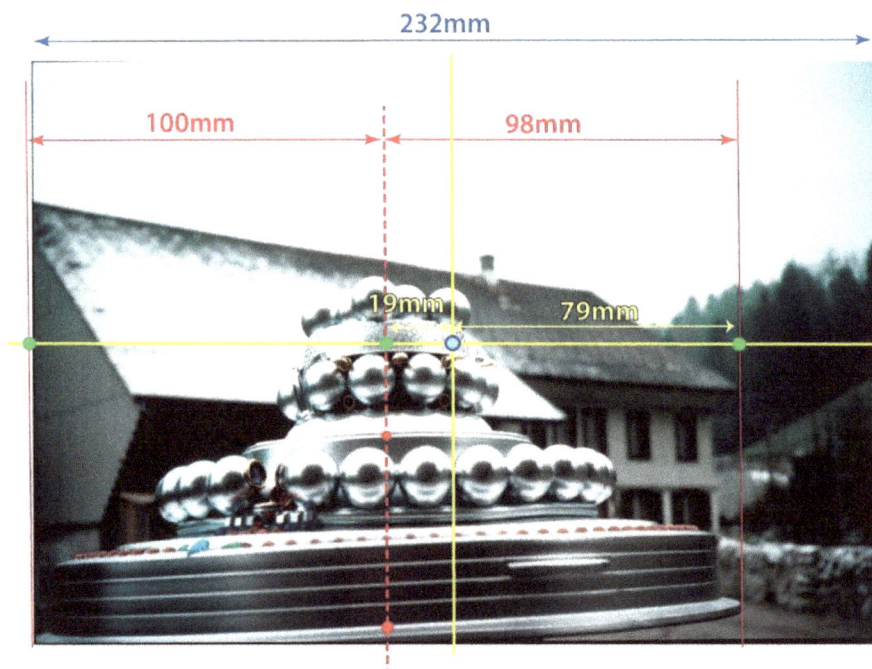

Figure 48- Measurements made on Photo #808 to determine the angle of view.

Angle of view

Figure 48 shows the WCUFO in Photo #808. Display this picture on a computer screen and measure its dimensions and distances. First, determine the photo's central axes and its center point (the yellow lines and blue dot in the figure). Next, draw the WCUFO central axis (red dashed line) and its parallel edges (red lines on each side). At this juncture, we notice and need to address a visual anomaly.

In Figure 48 the WCUFO seems to be asymmetrical. The central axis going through the WCUFO midpoints is not perpendicular to the WCUFO base. Why not? Is the WCUFO asymmetrical? We know from other photos that it *is* symmetrical. There is, however, noticeable distortion with this WCUFO base. Since the base is fine in other pictures, there are two possible causes for the distortion. One concerns electromagnetically induced distortion and the other optical distortion caused by the lens.

There were claims that this craft occasionally ruined photos due to electromagnetic frequencies it emitted, distorting color and light in most of the WCUFO with trailer shots, and here perhaps distorting the shape, like heat waves "bending" lines. A Christian Frehner 24 November 2009 email to Lock explained, "Quetzal said that the negatives and slides could also be negatively influenced, because the strong radiation of the ship would contort the photos if they were taken at too close distance. Therefore, the contours/outlines of the ship would show wavy lines and appear asymmetric, because the air would shimmer and vibrate just as in hot air." Temporary electromagnetic frequencies emitted from the WCUFO base might explain its distortion, as might emissions from the spheres which Ptaah is said to have called "swinging-wave accumulators" (See Contact Report #442 pp. 134-135.) This sphere designation suggests they accumulate electrical frequencies or resonances which might cause optical distortion in their vicinity.

The second possible cause is optical lens distortion. The most familiar forms of these are the *barrel effect*, the *pincushion effect*, and the *mustache effect*. We leave you to investigate and test these for yourself. Information on them is readily available online (Plumridge 2016). Photoshop and other software can now counteract these effects, although Ricoh-related correctional material is scarce. We have been able to largely counteract this photo's distortion using a pincushion effect correcting software. So we currently lean towards finding the pincushion effect largely responsible if the distortion is caused optically by the lens.

The doubt here lies in the fact that the pincushion effect usually occurs with telephoto lenses, which start at 70 mm, and Meier's Ricoh is a bit short at 55 mm classified as "standard" (35-70 mm). The distortion in Photo #808, however, does largely fit the pincushion pattern (Plumridge "What is

Pincushion Distortion?" 2016). The effect occurs more frequently with cheap cameras and Meier's was cheapish, and with large zoom lenses, but Meier's lens is just 15 degrees short of standard telephoto (70-135 mm).

Whichever is causing the distortion, electromagnetic frequencies or optical lens distortion, or both, we suggest measuring in the picture center rather than the edges. Ignoring the left side of the WCUFO and measuring only from the right side should increase the accuracy of calculations.

Having plotted the red lines, next measure the distances on the yellow horizontal axis line from the WCUFO central axis, and its right edge, to the center of the photo, the green points in Figure 48. We ignore the WCUFO tilt as it has little effect on our angle and size estimates. Measuring on the computer screen gives 79 mm from the picture center to its right edge and 19 mm from the picture center to the WCUFO axis. You can use these values to estimate the angle of view or make your lines and calculations.

Then, with a big sheet of paper, conduct the following four steps illustrated in Figures 49 and 50. Note the big sheet of paper gives camera values ten times larger than the real values, and a letter-size piece of paper gives the actual camera measurements at scale 1-1.

Step 1: Draw a triangle representing the camera. To understand the camera geometry, refer to "making a model of Meier's camera" in Experiment 5. We know Meier's Ricoh camera had a focal length of 55 mm, and the negative film width was 35 mm. For accuracy use a large 10 to 1 scale, necessitating a triangle representing the camera with a base of 350 mm on the big sheet of paper shown by the vertical segment at the right in Figure 49 top (35 mm for a small letter-size piece of paper). The other dimension, the focal length, is measured from this segment, towards the left a distance of 550 mm (55 mm on a letter size piece of paper). The result is an isosceles triangle as in Figure 49 top. Then draw a dashed line to the right, extending the central axis of the camera view.

Step 2: The distances in Figure 48 from the right edge, and the central axis of the WCUFO, towards the center point of the picture, are 79 mm and 19 mm respectively. However, we cannot directly use these computer screen values because the film width is only 35 mm while it is 232 mm on the PC screen. So the computer sizes of 19 mm and 79 mm need to be converted to film values.

Making these conversions is easy enough. Divide the real width of the negative film, 35 mm, by the width on the computer screen, 232 mm, to obtain a PC conversion factor:

$$PC\ conversion\ factor = \frac{35}{232} = 0.151$$

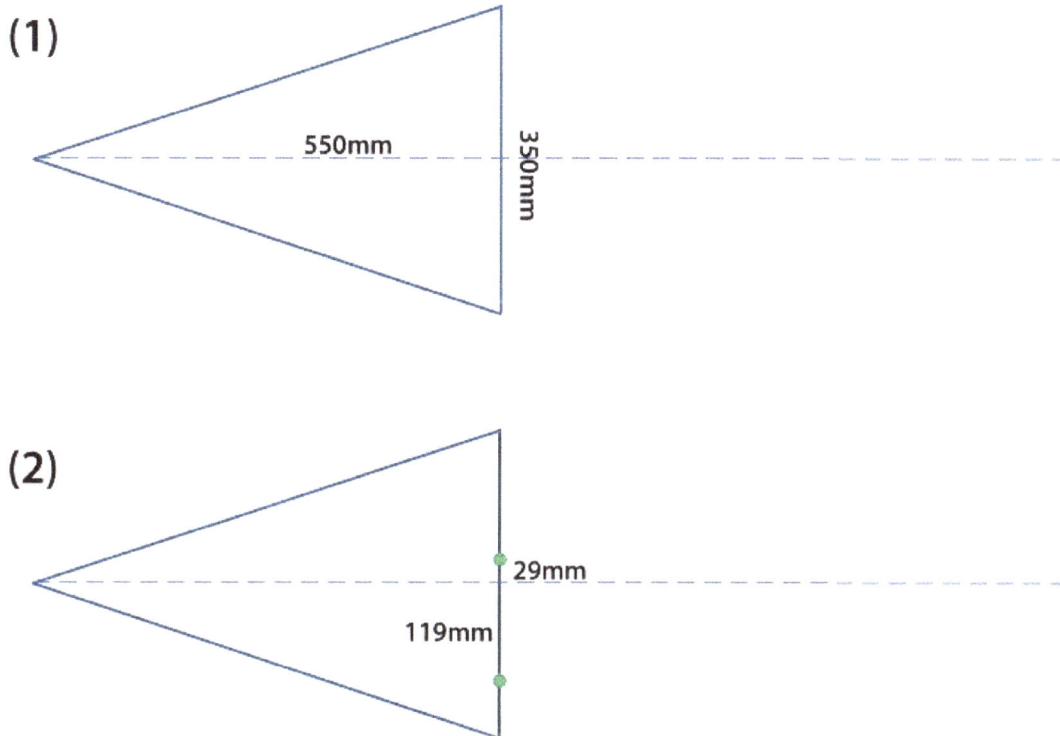

(1)

550mm

350mm

(2)

29mm

119mm

Figure 49- Steps 1 and 2 to estimate the WCUFO angle of view in Photo #808.

Multiplying any measurement on the computer screen by this factor gives real values on the negative film. So the 19 mm times 0.151 (PC factor) equals 2.9 mm, and 79 mm times 0.151 equals 11.9 mm. So we draw these values ten times bigger on the big sheet of paper: 29 mm and 119 mm (see Figure 49 bottom). Of course, at scale 1-1 on a letter-size piece of paper, they would be 2.9 mm and 11.9 mm.

Step 3: Draw lines from the camera origin (left vertex of the triangle) through and beyond the two points drawn in step 2. One line, the baseline, passes through the center of the WCUFO and the other goes towards the right edge of the WCUFO. With a protractor, measure the angle between these lines. This angle is the semi-angle of view or half the full angle. For Photo #808 we obtain a semi-angle of 14 degrees (Figure 50 top).

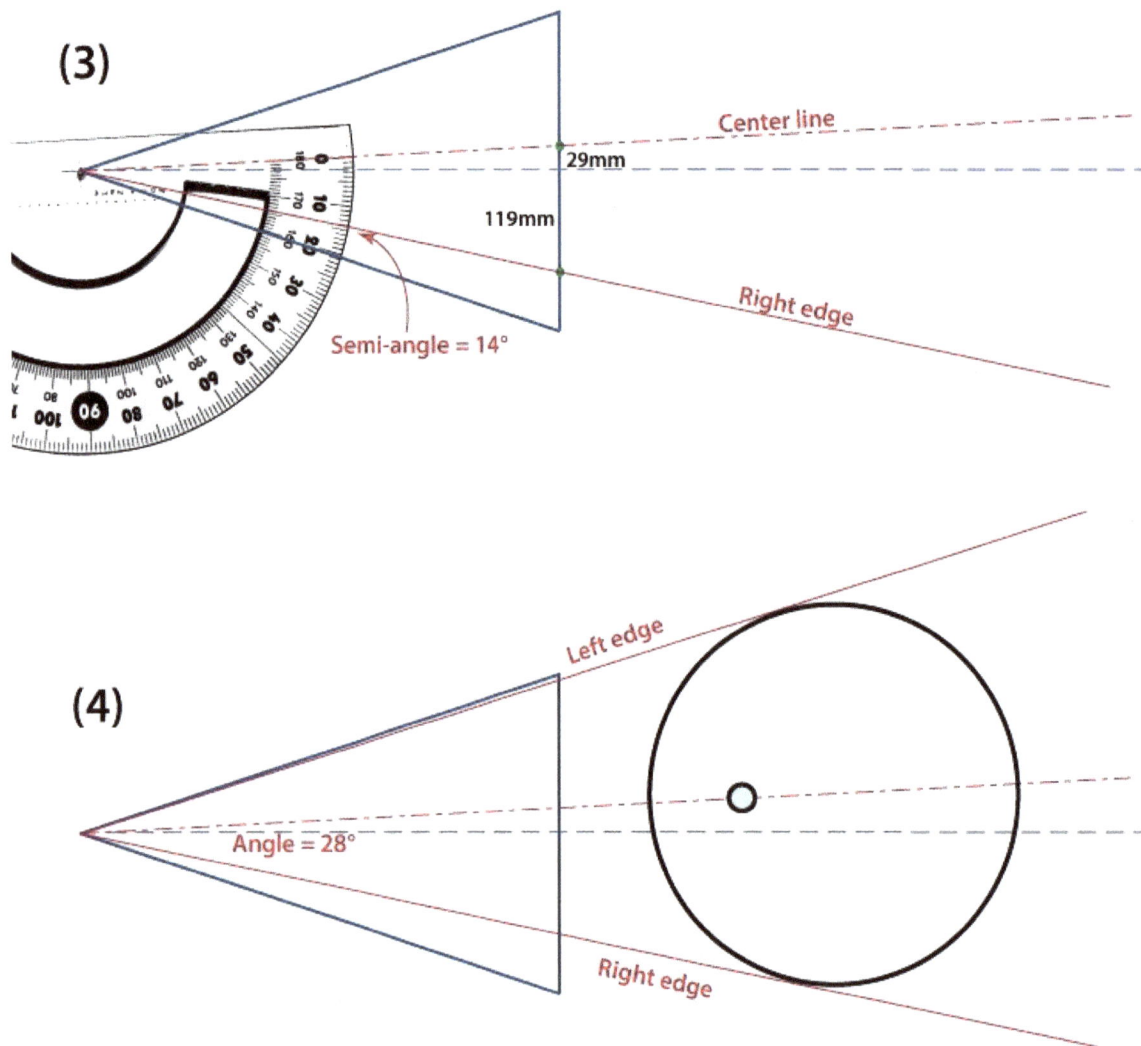

Figure 50- Steps 3 and 4 to estimate the WCUFO angle of view in Photo #808.

Step 4: Measure another 14 degrees with the protractor from the central axis towards the left side and draw the bounding line there for the WCUFO. We now have the full angle of view which is twice this 14-degree value, i.e. 28 degrees. Next, draw a black circle representing the WCUFO, as a reference (Figure 50 bottom). To know, however, if this is the real size of the WCUFO we need further work.

Upon completing the four steps, we know the WCUFO angle of view in Photo #808 is 28 degrees. For Photos #799 and #800 lower values result, because the WCUFO was farther away.

WCUFO size

Now we know the angle of view and the distance **Ds** from Meier to the nearest WCUFO sphere we can realistically estimate this WCUFO size.

Annex A charts the WCUFO proportions for both the 7 m and the 3.5 m diameter craft (Figure A1 and Table A1). We assume the proportions of the two craft are the same. There are very slight differences, and the big WCUFO can extend its central core upwards (see Experiment 11), but here we assume their proportions are the same. We found the lower tier of spheres located at 0.61 times the WCUFO radius measured from its center (Figure 51, top).

After performing the above four steps, on the same big sheet of paper, draw an angle of 28 degrees for Photo #808 as in Figure 51. This drawing is on a scale of 10 cm to one meter on Meier's property. On a letter-size piece paper, use a scale of 1 cm for each meter.

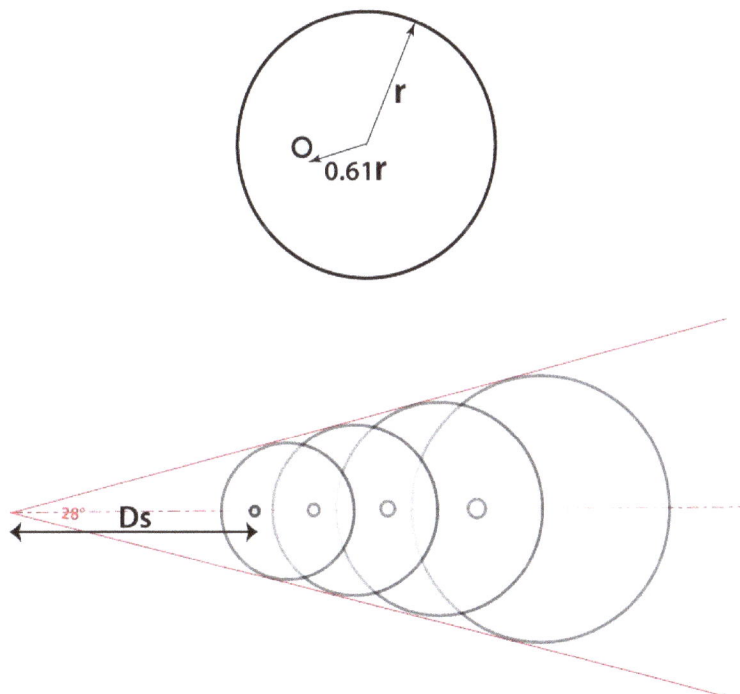

Figure 51- Several WCUFO sizes for Photo #808. The spheres at the lower tier are located at 0.61 times the WCUFO radius.

Use other pieces of paper to make several more WCUFO circular paper models, as shown in Figure 51. For example, WCUFOs of 0.5 m, 1.0 m, 2.0 m, 3.0 m, 3.5 m, 4.0 m, and any sizes you wish to test. A WCUFO of 3 meters is a paper circle of 30 cm, and a small half-meter diameter WCUFO model that skeptics claim Meier used is a little paper circle of 5 cm. After making the circles for different sizes of WCUFO draw on each one the nearest sphere at 0.61 times the radius, as shown in Figure 51. So, for a WCUFO of 2 meters in diameter, make a 20 cm diameter circle (10 cm radius). Its nearest sphere is then at 0.61 x 10 cm = 6.1 cm from the center. Finally, cut out all your drawn paper models.

Overlaying all your WCUFO paper cutout models on the drawing gives a layout like Figure 51 (bottom). Notice that the smaller the WCUFO, the closer the camera is to your little blue circle, the nearest sphere. You calculated **Ds** in a previous experiment, so you can now know how big this WCUFO is. Just measure **Ds** at scale, one meter to 10 centimeters, and find which WCUFO model fits best.

In your drawing, this best-fitting WCUFO paper model with its blue circle sphere closest to **Ds** is the one most closely representing the real size of this WCUFO.

Additional method to determine the WCUFO size

Knowing the angle of view, we now use some simple math and junior high school trigonometry to calculate the WCUFO size (its radius **r**) and the distance from the nearest sphere to the camera, **Ds**. In the process, we discover the relationship between the two.

In Annex A, you see the location of the lower circle of spheres to the WCUFO central axis is 0.61 times the radius. So the bottom tier of spheres is located at 61% of the WCUFO radius from its central axis.

In the example of Photo #808 the angle of view was 28 degrees, so the semi-angle is 14 degrees.

From Figure 52 and the trigonometric Sin = Opposite over the Hypotenuse, we obtain a simple formula that relates the WCUFO radius (**r**), the semi-angle (14 degrees), and the distance from the camera to the WCUFO center (**D**).

The method is:

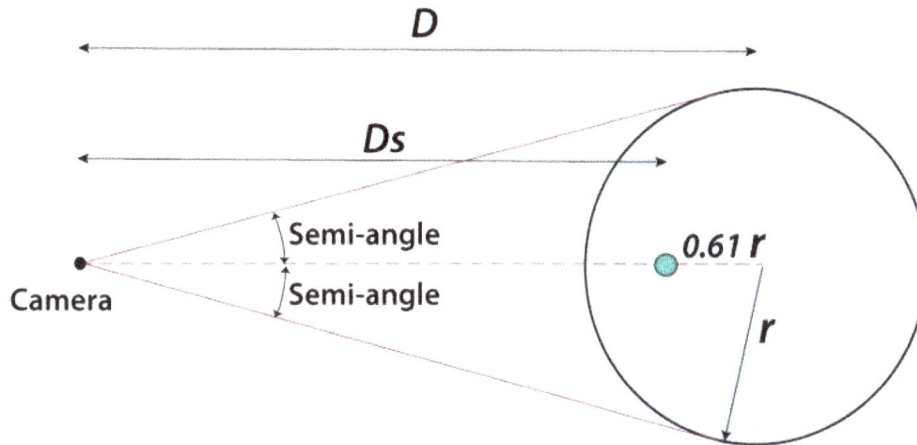

Figure 52- Basic diagram for WCUFO size and distance.

$$Sin\ (Semi\_angle) = \frac{r}{D}$$

$$r = D\ Sin(semi\_angle) \qquad (\textbf{r}\ \text{formula})$$

D is the distance to the center of the WCUFO and **Ds** is the distance to the nearest sphere. Also, from Figure 52 we know:

$$D = Ds + 0.61\ r \qquad \text{(Formula 1)}$$

Substituting Formula 1 in the **r** formula above:

$$r = (Ds + 0.61r)\ Sin\ (Semi\_angle)$$

$$r = Ds\ Sin(Semi\_angle) + r\ 0.61\ \ Sin(Semi\_angle)$$

$$r - r\ 0.61\ \ Sin(Semi\_angle) = Ds\ Sin\ (Semi\_angle)$$

$$r\,(1 - 0.61\,Sin(Semi\_angle)) = Ds\,Sin\,(Semi\_angle)$$

We obtain:

$$r = \frac{Ds\ \ Sin(Semi\_angle)}{1 - 0.61\ Sin(Semi\_angle)} \qquad \text{(Formula 2)}$$

where:

r   Is the WCUFO radius

Ds   Distance from camera to center of nearest sphere.

Semi-angle   is 14° for Photo #808.

With this formula, we can calculate the radius of the WCUFO, and twice this value is its diameter. All we need to know is the semi-angle of view (half the angle of view) and **Ds**.

With Formulas 1 and 2 we can construct Table 5. It tells us that the 3.5 m WCUFO would be at a distance of around 7.20 m (the closest sphere at 6.15 m) for #808. It also informs that any scale model of 55 centimeters as proposed by some skeptics must be centered at around 1.17 meters from the camera with the nearest sphere at 1 meter. However, we know that a UFO model at just one meter from the carriage house wall would display a huge reflection of this construction, and this is not what we observe in Meier's photos.

Also focusing at 1 m with a 55 mm lens causes middle and far distances to become very blurred, far more blurred than Meier's house and the distant trees in his picture. Check this with your SLR camera noting that the far end of Meier's house is about 38 meters away. Smartphones use high tech to give good depth of field even on zoom, so they are not suitable for reproducing Meier's SLR WCUFO shots, but they are okay for Experiment 4 since there you only need a good photo of your sphere reflection and not a reproduction of Meier's photograph.

Ds (to Sphere)	Photo #808		
	WCUFO radius	WCUFO Diameter	D (to center)
1.00	0.28	0.57	1.17
2.00	0.57	1.14	2.35
3.00	0.85	1.70	3.52
4.00	1.14	2.27	4.69
4.25	1.21	2.41	4.99
4.50	1.28	2.55	5.28
4.75	1.35	2.70	5.57
5.00	1.42	2.84	5.87
5.25	1.49	2.98	6.16
5.50	1.56	3.12	6.45
5.75	1.63	3.26	6.75
6.00	1.70	3.41	7.04
6.25	1.77	3.55	7.33
6.50	1.84	3.69	7.63
6.75	1.92	3.83	7.92
7.00	1.99	3.97	8.21

Table 5-   Possible WCUFO diameters in Photo #808, and distances from the camera to their central axis (**D**) based on the distance to the nearest WCUFO sphere (**Ds**). All values are in meters.

To find other WCUFO radii and sizes use Formulas 1 and 2 on pages 85-86 for photos like #799 and #800, or even #798 and #803. Other photos, however, present two complications: First, viewed from the camera, their WCUFO is towards the right, so the reflections show not just the north-east carriage house wall but also a portion of its side wall, necessitating a model of the entire construction. Second, we do not know the baseline direction in other photos because we cannot see how the WCUFO relates to Meier's home in the background due to extreme lack of home detail.

Suggested experiments

Based on the example presented here for Photo #808, we suggest you perform the following experiments:

A. Calculate the angle of view as before for Photo #799, or #800. If you have already calculated **Ds**, you can know the size of the WCUFO. Create a table similar to Table 5 for each photo in Annex D.

B. Having estimated well the WCUFO size in your previous experiments, you can use it to make a diagram showing the path the WCUFO traversed on the day when Meier took these eleven photos in his courtyard. For this, you need to calculate the angle of view in all the Figure 5 photos. Then, estimate the direction, the baseline, for each one, and do the reverse: estimate the distance to the camera once you know the size and the angle of view. Either the visual or mathematical method explained in this chapter suffice. You can then draw a scale map of Meier's property showing the subsequent movement of this object by plotting the sequentially taken photos. Figure 53 is our schematic of its movement. Anyone skilled in the use of 3D animation software could create an animated video of the WCUFO movement which lasted around one minute.

C. Use other photos, for instance, Photos #841 and #844 in Figure D3, Annex D. If this is a 3.5-meter WCUFO, how far away is it? Find the angle of view and estimate the distance if it is 3.5 meters in diameter.

D. In Figure 54, Erhard Lang conducts another test on Meier´s property. The photographer is very close to the carriage house wall. The camera is higher to look over the foreground vehicle. Lang is holding a 3.5-meter long pole. Compare Figure 53 with the overlay of two WCUFO photos in Figure 21, other Meier WCUFO photos in Figure 5, and figures in Annex D. Check the size of the pole Lang is holding against the background image of Meier's home. Is this 3.5-meter pole angle of view representative of where the WCUFO was flying? How big in this photo (Figure 54) would the WCUFO look if it was a 55 cm diameter model?

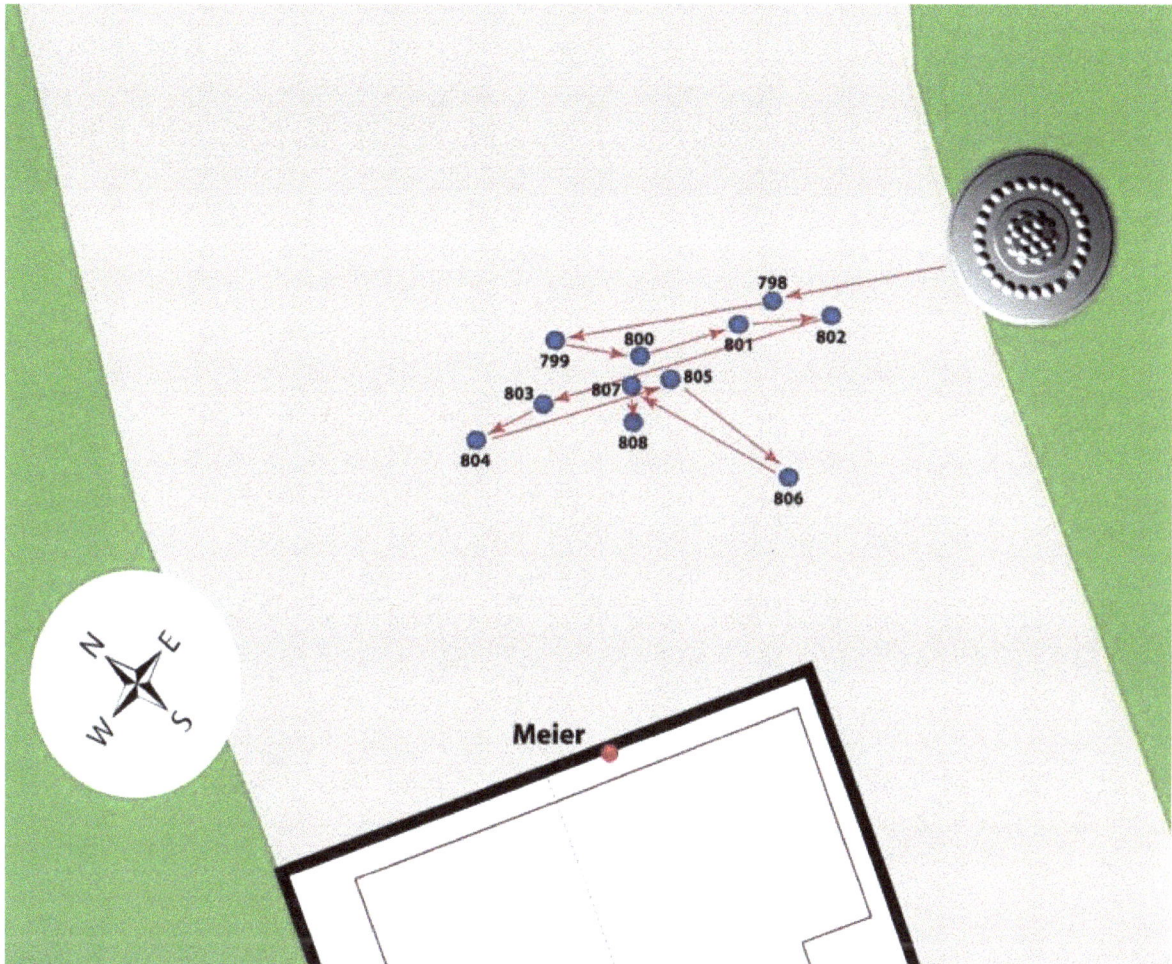

Figure 53- WCUFO movement above Meier's courtyard plotted from his 11 sequential photos. Blue dots indicate the approximate WCUFO center in each photo. Meier is proximal to the carriage house wall at the bottom. A 3.5 m diameter WCUFO is shown assuming a southeast arrival.

Figure 54- Testing the WCUFO location during Meier's photo session, Erhard Lang holds a 3.5-meter long pole in the SSSC parking lot (the courtyard).

Chapter 9

Experiment 7: Distance estimation using the camera formula

INTRODUCTION

In this experiment, we use a tool related to camera geometry called the *camera formula.* It shows the distance to an object when we know the object's size. For example, when viewing a tree in a photo, based on the age of this tree, we can estimate its height and then know how far away it was from the camera. In our case, looking at an object like a window on Meier's background home in the WCUFO photos in the courtyard, if we know the size of this window we can know how far away his camera was from this part of his house.

In Experiment 3, Chapter 5, we established that Meier's camera was very close to the carriage house wall, below the roof extension. We reached this conclusion by analyzing the optical alignments evident in his photos. At the end of that chapter we pointed out that the camera formula could help us to confirm this finding; in this experiment, we do this.

METHOD

The camera formula aka the **pinhole camera model** is a method by which we can know how far away any object is from the camera once we establish how big the object is and the camera characteristics. This formula states that *the ratio of the object's size on the sensor or negative and the size of the object in real life is the same as the ratio between the camera's focal length and the distance to the object* (Stack Exchange "Photography").

We express these four elements respectively as **h/H = f/D**

Looking at some of Meier's photos he took in his front yard, like #800, we notice several windows on the background wall of his home. For confirmation of the camera's location in the courtyard, we can use the calculation on Professor Jim Deardorff's website, where he used the window marked **W** in Figure 55. The height of this window was measured by Frehner to be 1.20 m (Deardorff *The Wedding-Cake UFOs*).

Figure 55- The height of a window (**W**) in Meier's home measured on the computer screen and used to find the camera's distance from the window.

The measurements for Figure 55, on a computer screen, were:

- Width of photo on the PC screen = 324.5 mm
- Height of the window opening = 16.5 mm (±5%)

This height of 16.5 mm is on the computer screen. We need to convert it to the equivalent 35 mm negative measurement. To do this we need a PC factor expressing the ratio between the 35 mm width of the negative to the 324.5 mm measured on the computer monitor:

$$PC\ factor\ =\ \frac{35\ mm}{324.5\ mm}\ =\ 0.1079$$

If we multiply every measurement we make on the computer monitor by this factor, we convert it to this equivalent negative film measurement. So the height of the window measured in the negative film **h** is:

$$h = 0.1079 \text{ x } 16.5 \, mm = 1.78 \, mm$$

Height of window opening on the 35 mm film = 1.78 mm

From the camera formula:

$$h/H = f/D$$

$$\boldsymbol{D = H \ (f/h)}$$

Where:

   D = Distance from camera to object in question;

   f = focal length of camera = 55 mm;

   h = height of object's image on the 35 mm film (1.78 mm)

   H = height of actual object = 1.20 m (window opening).

Therefore:

$$\textbf{D = 37 m (\pm 5\%)}$$

So the distance from the camera to the window **W** in Figure 55 is in the range of 35.2 m to 38.9 m. In the plan view of Meier's residences (Annex fig. B6), we found this distance to be 36 m.

At this distance, the only conclusion possible is that Meier and his camera were very close to the carriage house north-east wall when he took his photos in the courtyard.

You may note that the camera formula works with any object of known size, whether its orientation is vertical, horizontal or diagonal, but it has to be in front view. In the case of the window the height length is in front view, but the base of the same window is not because one extreme of the window base is closer to the camera than the other. Alternatively, we could use the vertical size of the photo too, in which case we would use the vertical 35 mm film size (23.3 mm). Either way, the results come out the same.

Suggested experiments

Now you understand the camera formula we suggest the following calculations:

A. For Photo #841, Annex Figure D3 (top), if the tree is 7 meters high, as Meier recently confirmed to Frehner, how far away was Meier when he took this picture? (Meier confirmed the 15-meter height for this tree cited in *Photo-Inventarium* page 111 is incorrect.)

B. For Photo #844, Annex D, Figure D3 (bottom), if the WCUFO is 3.5 meters in diameter, how far away from the WCUFO was Billy Meier when he took this photo?

C. In Figures 56 and 57 Photo #834: If this is a 3.5 m WCUFO how far away was it from the camera? If it is a 0.55 m model how far away was it? If it is very close, would the tree between the camera and the UFO be visible in the sphere reflections? If so, how big would it be? (Suggestion: use the left side of the WCUFO for the semi-angle).

Figure 56- Photo #834 of a 3.5 m WCUFO. Meier, flying on a 7-meter WCUFO at treetop elevation, took this picture through the branches. April 3, 1981, at 1:10 pm. (Large format photo in Annex D.)

Chapter 10

Experiment 8: Mapping the WCUFO's local environs

INTRODUCTION

The WCUFO presents an interesting opportunity to verify its surroundings by checking the reflections in its spheres. Its spheres reflect images of the vicinity, and in this Experiment 8, we explain a simple graphic method to map in detail the WCUFO's local environmental surroundings by charting its reflected sphere images. Later in Experiment 10 we see they can be used to produce stereoscopic 3D images.

If we see the reflection of a group of trees, even in a very blurred image, there is an analysis we can do to estimate the relative position of some of the trees to the WCUFO. To an observer in or on the WCUFO, how tall would the trees appear to be? What elevation, in degrees above the horizon, is the top of the reflected tree? This angle can give us some detail of the surroundings at the precise moment of the photo shooting. In this experiment, we look at reflection details in Photos #834 and #844. Finally, in this experiment, we make an illustrative diagram of the WCUFO environs in Photo #844.

Meier shot Photo #834 (Figure 57) at 1:10 pm on April 3, 1981, around 20 minutes before shooting Photos #841 and #844. It shows a WCUFO in daylight from behind trees. Meier says he was on another 7-meter WCUFO flying gently at treetop level taking pictures of this smaller 3.5-meter diameter WCUFO. Our analysis shows this is indeed an object of around 3.5 meters in diameter, just as those in photos #841 and #844 (analyzed in Experiment 9 Exercise 3).

If you have not performed Suggested Experiment C in Experiment 7 consider the following: If Figure 57 shows a little WCUFO model half a meter in diameter instead of a real WCUFO, it must be at around 3 meters from the camera. If we imagine a small model of this size and a tree between the camera and the model, how big must the tree be in the sphere reflections? If, on the other hand, it is a distant 3.5-meter diameter WCUFO at around 20 meters from the camera, how big must the tree reflections be?

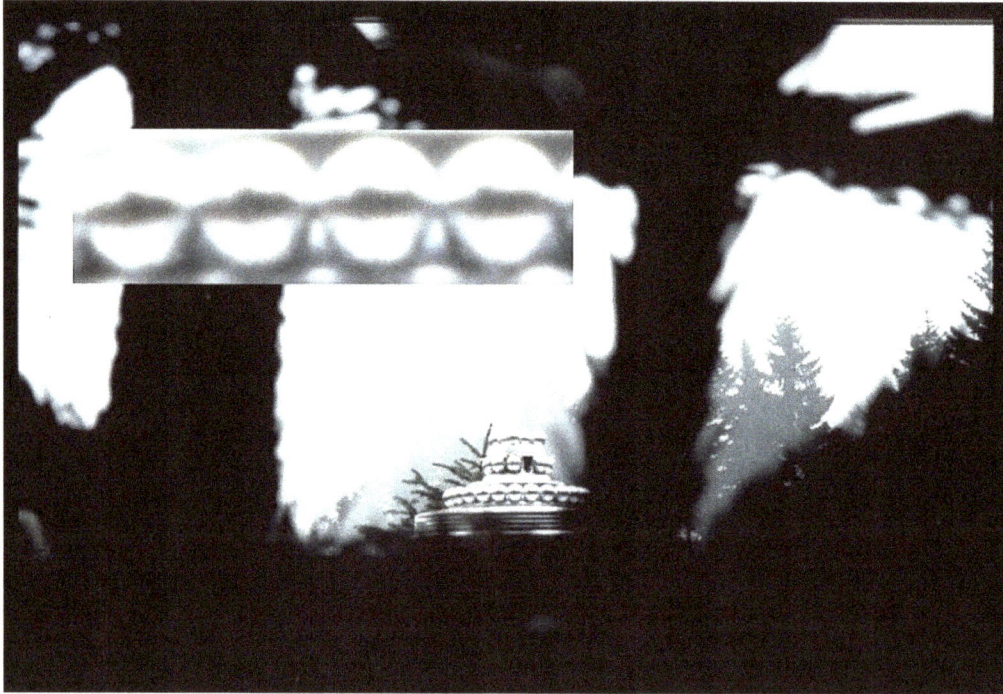

Figure 57- Photo #834 of a 3.5 m WCUFO. Meier says he was on a 7-meter WCUFO at treetop elevation taking this photo through the branches. April 3, 1981, at 1:10 pm. (Large format photo in Annex D.)

MATERIALS

For this experiment you require:

- Templates as illustrated in Figures 58 and 61. You can get downloads here:

  www.rhalzahi.com/docs/GB-Reflections-template.pdf

- A ruler to measure distances in the photos, so the location of any specifically reflected object can be transposed onto the template.

- A protractor, two straws, and a pin to make a simple angle measuring device.

METHOD

Sphere reflections: locating reflected objects

To estimate the direction of an object reflected in a WCUFO sphere we plot the object on a Sphere Template (Figure 58). First, you need to copy Figure 58 or download the template from the "Materials" link above. Then check the images from a photo you have selected, and finally draw to scale the objects' locations on the template. Here is how you do it.

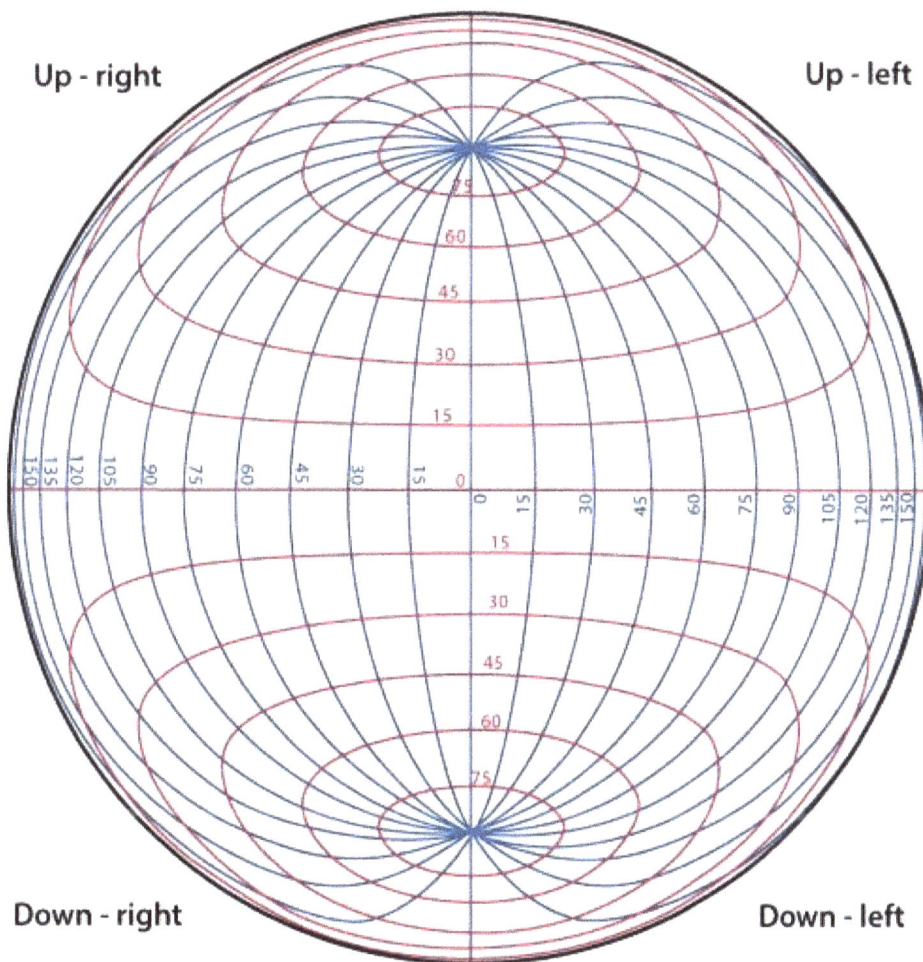

Figure 58- Sphere Template for plotting objects vertically and horizontally.

Plotting objects vertically & horizontally

On the Sphere Template in Figure 58, blue curved vertical lines indicate the horizontal angle and red curved horizontal lines indicate the vertical angle. The angles are the numbers on the Sphere Template. The center of the Sphere Template is, of course, 0 degrees vertically and horizontally. We need to define specific points of the object, for example, the tree's top and base. With one or more points marked on the template, check the template red vertical and blue horizontal angles as viewed on the analyzed sphere. Check between where the points fall on the vertical or horizontal template lines and draw the objects. The following examples clarify this process.

A hypothetical example

Figure 59 shows how to plot onto the Sphere Template an object reflected in a sphere. Our object reflected is a tree. Assume that from the base of a tree you observe a reflecting sphere in front of you (Figure 59 left). The elevation of the tree seen from the sphere is 60 degrees. The reflected tree image seen in the sphere goes from the exact sphere center up to a mid-point on the Sphere Template's central vertical radius (Figure 59 center). So on the Sphere Template, we draw the corresponding tree, as seen in the sphere reflection, i.e. from the Sphere Template center half way up its central radius, where it tops at the 60-degree horizontal red line (Figure 59 right). In a similar manner, plot the width of the tree's bottom branches, as you see them in the sphere image, which here corresponds to approximately 16 degrees right and 18 degrees left of the Sphere Template's central vertical axis (Figure 59 right).

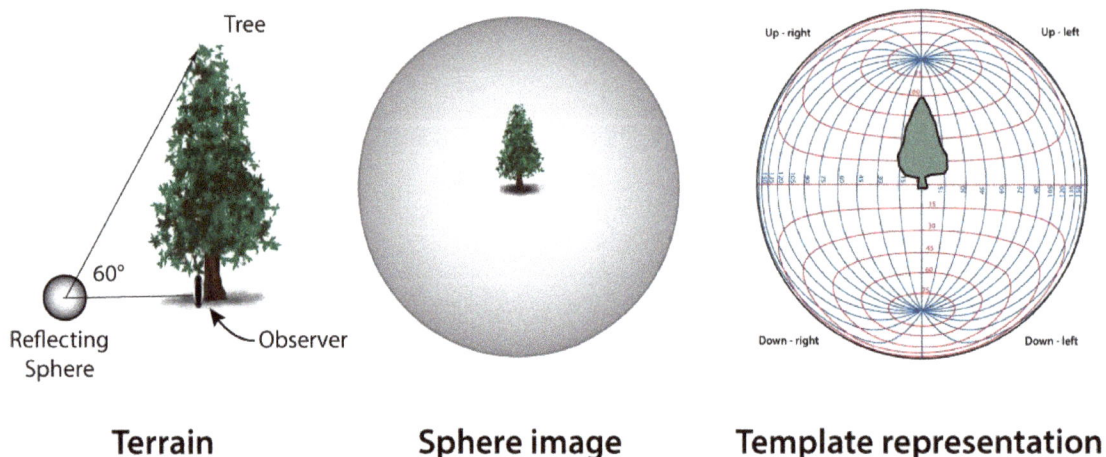

Figure 59- Plotting an object vertically and horizontally onto the Sphere Template.

The baseline and the base plane

Now, in Figure 60, we evaluate a reflection in the WCUFO sphere colored red. Camera "C" (the pyramid symbol) takes the photo. The red dotted line connecting our red sphere to the camera is the baseline. Measure two angles, one horizontal (**H**), and one vertical (**V**) from the sphere to the top of the tree reflected in the sphere (Figure 60). Viewed from the sphere, the horizontal angle **H** measures from the red dotted baseline to the left of the camera, and the vertical angle **V** is above the plane of the baseline. We know the direction, but we do not know how far away the tree is from the WCUFO.

Notice the angles are always measured from the WCUFO sphere. Also, the baseline linking the sphere and camera is not always horizontal; in the WCUFO photographs, Meier's camera is at times above, below or at the same level. So there is a base plane containing the baseline, and it is not always horizontal. Always measure vertical angles from this base plane. Here, we assume the base plane is horizontal, which provides an easy example for accurately mapping the WCUFO's surrounding objects.

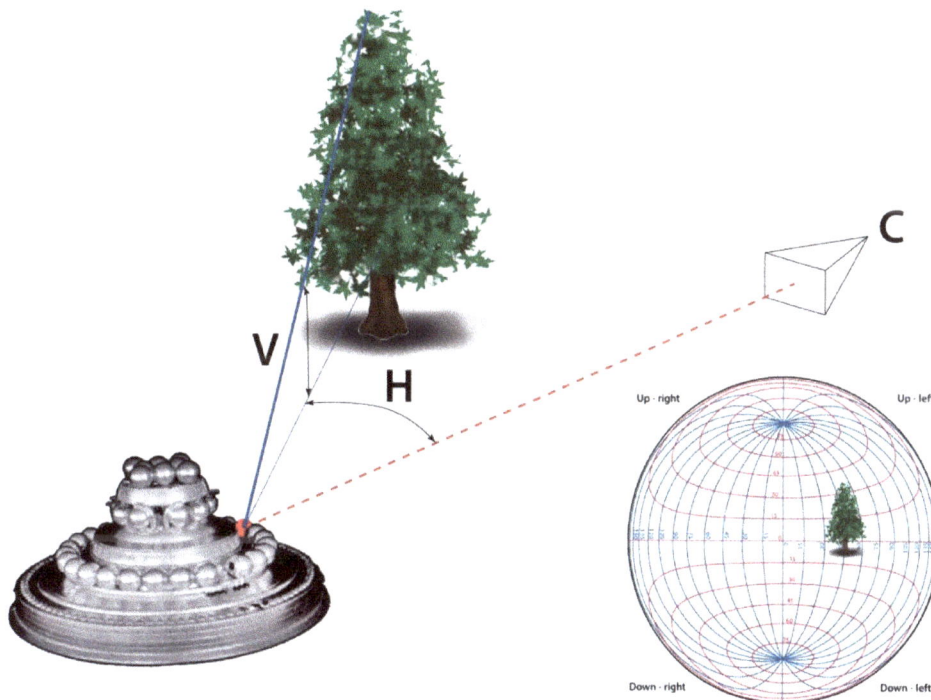

Figure 60- Horizontal and vertical angles **H** and **V** measured from the red sphere. From the sphere's viewpoint, the tree top is to the camera's left above the base plane with its red dotted baseline linking the sphere and camera.

Figure 60 (bottom right) shows the Sphere Template and tree reflection seen from the camera. In this case, treating the WCUFO red sphere as the Sphere Template, the horizontal angle **H** of the base of the tree is around 45° to the left, and the vertical angle **V** of the treetop is 37° above the base plane.

Cylinder reflections

The base of the WCUFO has a cylinder-like edge. Some bright reflections like sunlight, and dark reflections like a tree or the carriage house, are discernible on its surface. For cylinder reflections, we use the Cylinder Template in Figure 61 (found with the Sphere Template download link in the "Materials" section above). With this Cylinder Template, we only measure the horizontal angle because there is no vertical angle.

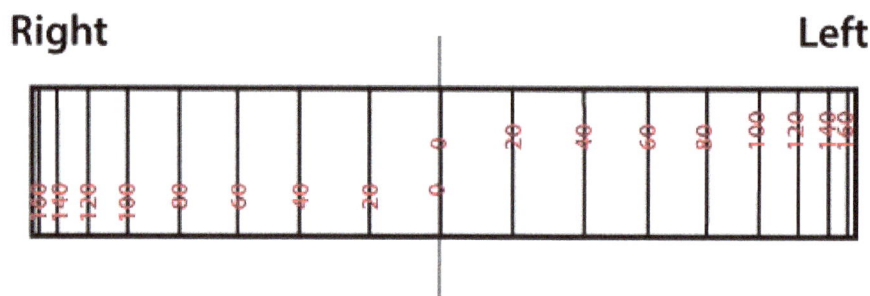

Figure 61- Cylinder Template for measuring horizontal angles only.

Angle estimations

We can estimate what the template angles represent in reality. For example, when finding a sphere reflection treetop at 30° above the horizon, or above the base plane, how high is this in fact? Going outdoors and pacing out a distance from a nearby tree until its top is at 30° can give an idea of how the WCUFO environs looked. Similarly, if a reflected object's location is at 135 degrees, horizontally towards the left, what does that mean?

There are two simple ways to make angle estimations. The first and easiest one is to use your hands. Extending your arm and opening out the hand gives an angle between the index finger and thumb tips of approximately 15 degrees (see Figure 62).

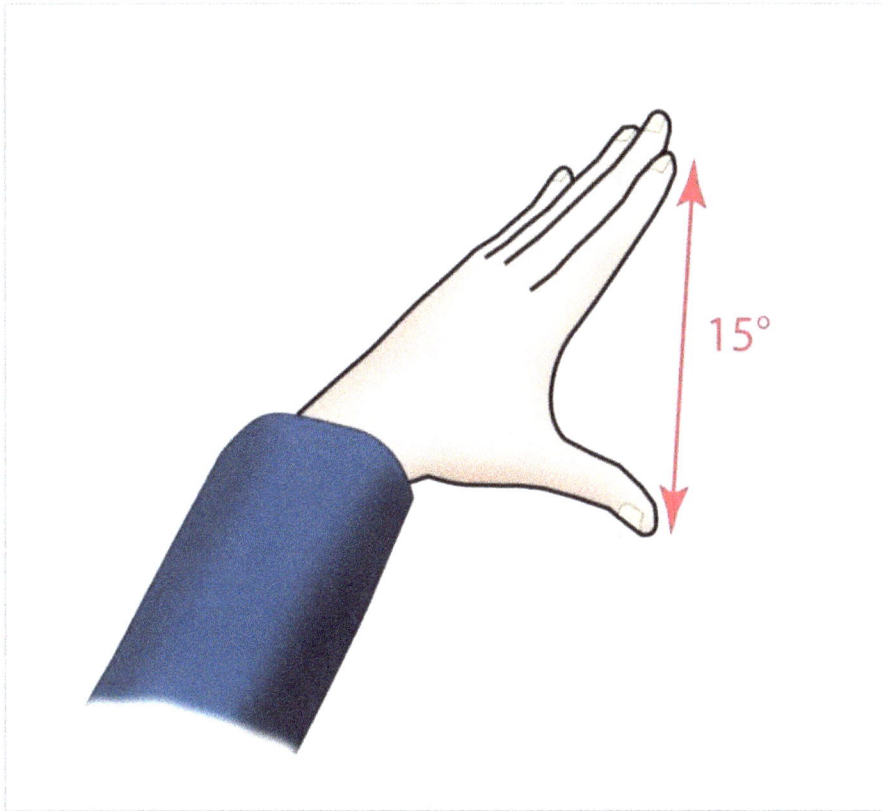

Figure 62- An easy way to estimate angles: Extend the arm, open the hand, and separate the thumb and index finger. The distance between their extremities is about 15 degrees.

Using the hand in this way is an easy rule of thumb method to estimate angles. To measure an elevation of 30 degrees above the horizon, use twice the measure of 15 degrees indicated in Figure 62. You can also use fractions. Just remember to extend the arm.

Another approximate method, though probably more precise consists in constructing an angle measuring device as shown in Figure 63. To make it, place two straws each side of a protractor and pin them together in the center. We can look through the straws to locate objects so they can be useful. Fix one of the straws on the zero degrees mark. The other one should rotate. Figure 63 shows the device measuring an angle of 37 degrees.

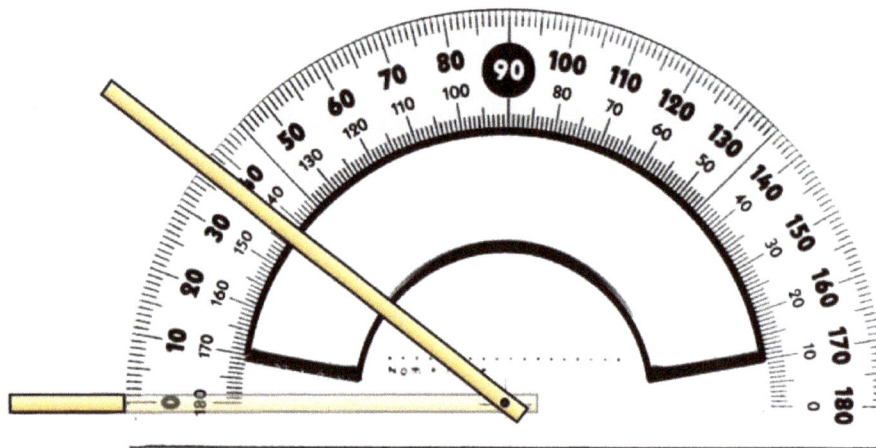

Figure 63- A simple angle measuring device made with a protractor and two straws. Place the straws either side of the protractor. Fix one at the zero degrees mark. The other rotates around a pin inserted in the protractor center base to join both straws.

Two suggested experiments

A- Photo #844 analysis

Now, using both the Sphere Template and Cylinder Template, you can finally draw a diagram to illustrate the WCUFO scene in Photo #844. In Figure 64 we see this WCUFO casts a huge shadow over the Norway Spruce, and later in Experiment 9, Exercise 3, we confirm the WCUFO is proximal to the tree. Now, however, using the available template, chart vertical and horizontal positions of reflections in the sphere circled red in Figure 64. Next, with their data entered on the template, plot the reflected objects' geographic locations.

Figure 65 is an enhanced and zoomed image of the red sphere in Figure 64. Increasing the contrast (Figure 65 left) we see a dark green triangle, which is the reflection of a tree. A light brown section to its right is a more distant group of trees. Darkening the image (Figure 65 right) the sun's reflection and shape become clear in its top right section. The brightness below the sun's reflection is the sun rays reflecting off the shiny WCUFO base. The dark image on the sphere's right side is the same tree reflected twice: in the sphere under study, and the contiguous sphere to its right. Use the triangular dark green image and the sun's reflection to estimate the positions of the tree and sun as seen from the sphere.

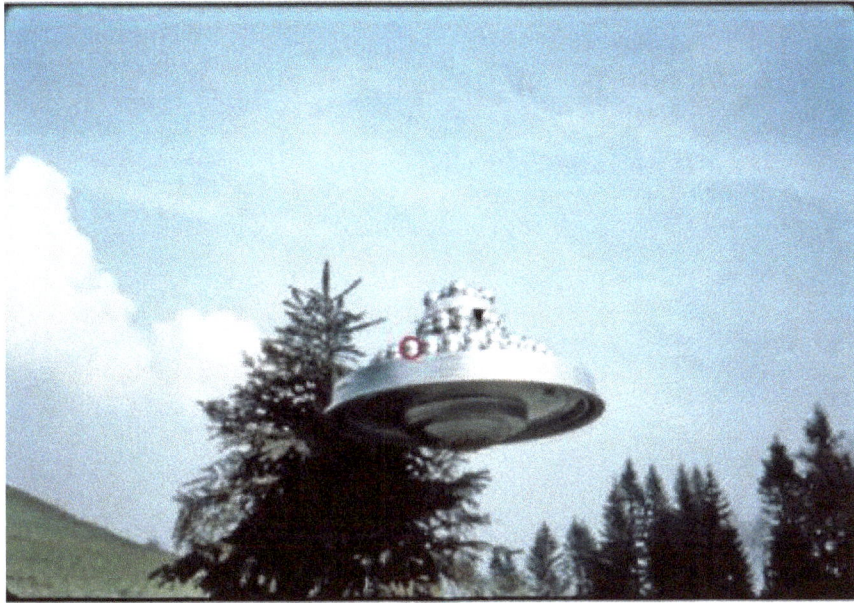

Figure 64- Photo #844. A WCUFO casting a huge shadow over a nearby Norway Spruce. The red circle highlights the sphere under study.

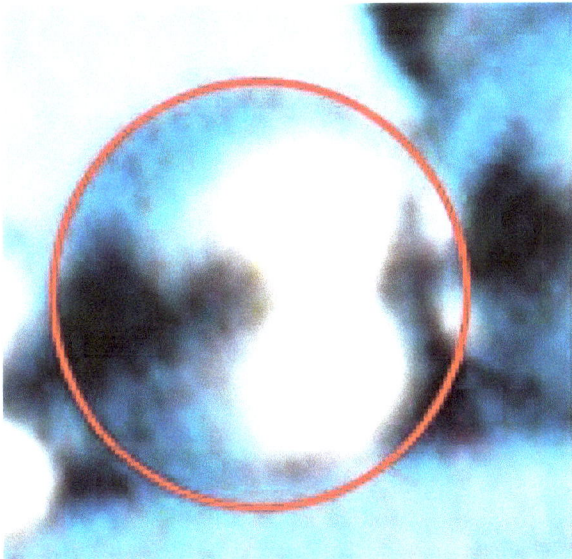

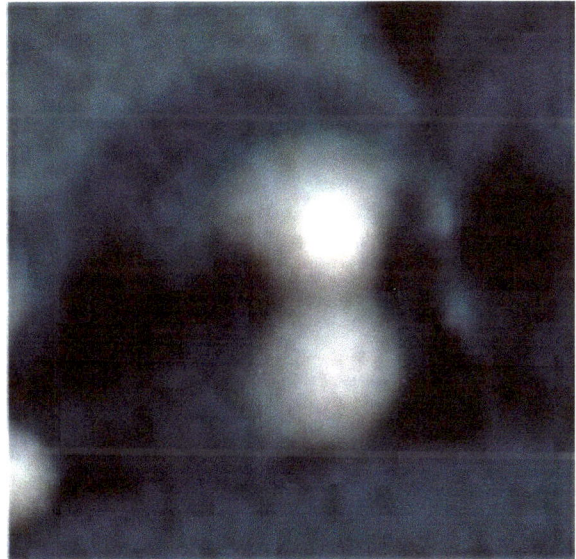

Figure 65- The analyzed sphere. Left: Increasing the contrast the nearby tree is visible as a dark green triangular shape. Right: The same sphere darkened reveals the sun's reflection as a white spot.

Now, transpose these observed images in Figure 65 to the Sphere Template (Figure 58, or the downloaded template), being careful to maintain the same proportions.

Also, use the Cylinder Template and transpose the sun's position based on what you have in Figure 66.

Figure 66- Darkened image of Photo #844 WCUFO. The sun's reflection is made visible as little white lines. The vertical yellow line marks its position.

Answer the following questions:

- What are the horizontal and vertical angles of the treetop? Imagine that you are at the WCUFO red sphere looking out into the environs. Use either of the angle-measuring methods, your hands or the simple angle measuring device. If Meier is some meters away in front of you, where is the treetop located? Is it on your left or right? Is it in front or behind you?

- Find the sun's position. What are its horizontal and vertical angles? Is the horizontal Spherical Template angle about the same as the Cylinder Template horizontal angle? Imagine again being located at the sphere under study. If Meier and his camera are in front of you, where is the sun? Is it to the left or the right of Meier? If it is around 1:35 pm, April 3, 1981, which direction is north? Was Meier towards the north, south, east or west from the WCUFO?

- If this WCUFO measured around 3.5 meters in diameter, using the angle of view method from Experiment 6, or the camera formula from Experiment 7, estimate the distance from this WCUFO to Meier's camera.

- Plot and draw an illustrative diagram or map, like Figure 67. Locate Meier at scale. Draw the direction of the sun rays and approximate the location of north. Use a ruler and protractor.

Figure 67- Illustrative diagram plotting the tree, WCUFO, and Billy Meier.

B- Photo #834 analysis

Figure 57 shows Photo #834 and the inset displays the sphere reflections. An enlargement of these sphere reflections is in Figure 80. Figure 68 shows an enhanced image of a WCUFO sphere reflection in #834. Red arrows indicate some of the reflected trees.

Make your analysis of Photo #834 using Figure 68 and answer the following questions:

- Based on Figure 68, how many main trees do you estimate there are in the reflected image?

- Analyze the tallest tree in the reflections. Use the template in Figure 58. What is the angle of elevation of this treetop?

- Imagine you are in this WCUFO watching Meier behind the trees. Where would he be? (Remember the sphere reflection rules.)

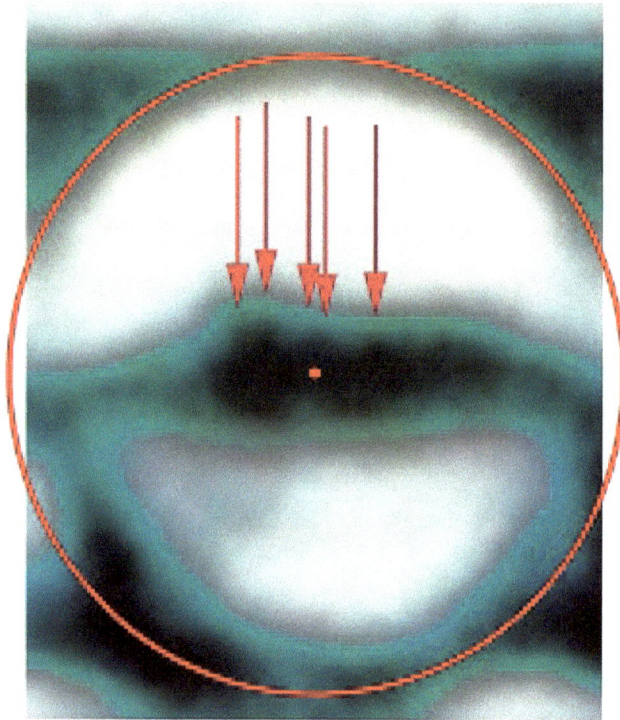

Figure 68- Photo #834 enlarged sphere reflection enhanced to show reflected trees appointed by red arrows. Trees farther away are not given arrows. The red dot is the center of the sphere, and Meier's camera must be there behind, between, or in a central tree under an arrow.

- From your viewpoint in the WCUFO find an object around you, like a tree, of about the same elevation. How tall does it look viewed from the WCUFO? Use any of the methods practiced to estimate angles.

- Use the camera formula from Experiment 7 to calculate the distance to this WCUFO assuming it is 3.5 m in diameter.

- If this WCUFO is a small half meter diameter model, using the camera formula, how far away would it be? What must the treetop elevation be from the camera and this model? Make a rough estimate and compare it with your previous estimate on the Sphere Template. Do your findings show this WCUFO is a small model or a distant bigger object?

- Make a diagram of the surrounding trees with the WCUFO in the middle. Just estimate, since you may know the direction and elevation of each tree but not their distance from the WCUFO. You can assume the distance to each tree relates to how dark or light, or how big or small the tree is in the sphere reflection – we assume the trees are all approximately the same age, and therefore the same basic height. In real life, they appear to be so at this site.

Make similar analyses using other photos. For example in Photo #808, plot the elevation of the carriage house top vertex, based on the reflected image in the nearest sphere. Alternatively, find other WCUFO photos and perform more tests using the templates and methods you acquired and carried out in this chapter.

Be aware that the measurements of angles beyond 150 degrees and close to the extremes of the Sphere Template and the Cylinder Template become increasingly inaccurate and questionable since a little variation in position can make a big difference in the resulting angle.

Chapter 11

Experiment 9: Image processing
photo images to discover hidden details

INTRODUCTION

In this experiment, through image processing, you analyze available WCUFO photo images to reveal their hidden details, and again, critically analyze these pictures by using your findings to answer questions.

Image processing is very simple today. In the 1970s and 80s, when Billy Meier took his photographs and videos, it was by no means an easy task. It was only possible in the most sophisticated labs. Watching the classic movie *Contact (from the Pleiades),* the product of Lee and Britt Elders investigations, we see that to scan one negative of one Meier photo took a specialized lab one full week. Today it takes just a few seconds with a piece of hardware purchased at the nearest electronics store.

This experiment consists of four simple image processing experiments or exercises. We highly recommend involving yourself in this simple image processing world which reveals remarkable hidden details within some of Meier's photos. We encourage and assist in finding suitable software and pictures you can analyze. Then you become the researcher making findings.

MATERIALS

Finding the right software and pictures

The computer tool Photoshop is good for editing images and enhancing hidden details. If you are familiar with it or have it, we suggest using it.

If, however, this tool is too expensive where you live, you can download and use for free (so-called "Free Software") a tool called GIMP (GNU Image Manipulation Program) available at https://www.gimp.org. How-to-use

Manuals can also be downloaded. It is an excellent tool for performing the exercises in this experiment. It can be installed using "torrent" or a direct download of an executable file. Unfortunately, it is "not available for download" in some countries like Japan. Some countries, perhaps for commercial or spyware reasons, block or do not permit its downloading. Even in these countries, though, you can often find extra "brightness" and "contrast" functions already installed on your computer by a camera company or the PC manufacturer.

In Japan, for example, a new Toshiba PC of around 2015, includes *Adobe Photoshop Express Toshiba Version Only* ready installed. This program is a basic Photoshop but has enough functions and capability ("brightness," "contrast," "highlight," "filter," and more) to perform the following experiments and obtain most of the very revealing details. So if you do not have Photoshop and cannot afford it, first search your computer's programs because today a decent one may already be there. These programs are usually for cropping and editing personal photos.

If your computer has a strong image processing program and you have tried image processing these Meier pictures before without success, it was probably the multi-generational, poor quality, or low-resolution of the pictures used that was the problem. There are now many versions of these pictures available, but it is the ones of early generation and highest resolution that reveal details that have remained hidden until now. The ones linked to in these four experiments are the high-resolution pictures, so if you have not had success to date, download and try the images in this experiment. Exercise 2 Photo #850 is even a high-resolution wallpaper download, and all four make acceptable to great wallpapers once downloaded.

If you lack skill with most available software, ask a friend or relative for assistance.

A scientist and researcher definitively asks and answers questions. So in the following four exercises, you do this while processing the images.

METHOD

Exercise 1: Photo #873

Go to: http://www.rhalzahi.com/images/Pic-873.jpg

Image:	Do:
	Download the image. Vary the brightness and contrast a few times. Due to contrast distortions at high levels, it is better to concentrate on brightness and use contrast minimally. Alternatively, use the tools "curves," "levels" or "Shadows/Illuminations."
	Something should be visible below the WCUFO using only "brightness." In other countries also try their versions of "filter," "light," or "highlight."

Findings from questions:

Do you see anything below the WCUFO?

Can you see the halo around this object?

Where do you think this object was flying? Where must the photographer be?

Is there any known object below the WCUFO? If so, does it give an idea of how big this WCUFO might be?

Exercise 2: Photo #850

Go to:

http://www.futureofmankind.co.uk/Billy_Meier/File:Figu-wallpaper_f850_1920x1080.jpg

Image:	Do:
	Download the image. Vary the brightness and contrast a few times. Alternatively, use the tools "curves," "levels" or "Shadows/Illuminations." Again, try using more brightness than contrast, and in other countries try their versions of "filter," "light," or "highlight."

Findings from questions:

Do you see a tree very close to the WCUFO?

Do you think this is a real big tree or a miniature or bonsai? (Google or Bing search "bonsai images" if unsure.)

Is the tree in front of the WCUFO?

Does it give an idea of how big this WCUFO might be?

Exercise 3: Photos #841 and #844

Go to:

http://www.rhalzahi.com/docs/Pictures-Tree.zip

Image:	**Do:**
	Download the Zip file. Extract both photos. Enhance the brightness and contrast of each one, or use the tools "curves," "levels" or "Shadows/ Illuminations." In other countries also try their versions of "filter," "light," or "highlight."
	These photos are part of a series taken by Meier on April 3, 1981, at around 1:30 pm while he was walking towards the WCUFO hovering static in front of the tree in the picture.

Findings from questions:

Is it the same tree in both photos, or are there two different trees here?

Is this a real tree? Google search in google.com an image of a Norway Spruce tree and compare its branches and leaves. Check the size of the trunk.

Is the WCUFO close to the tree or is this a false perspective trick? (False perspective occurs with a little model very close to the camera and a distant tree in the background giving a false impression that both are close to each other). Qualify your answer.

Do you see the large shadow cast by the WCUFO on the tree branches? What does this shadow tell you?

How tall might this tree be? How big is this WCUFO?

Comparing the top right cloud configurations, hillside, and sky, with the bottom left, do they look similar? What do they tell you?

Exercise 4: Photo #808

Go to:

http://www.rhalzahi.com/images/Pic-808.jpg

Image:	Do:
	Download the image. Zoom different areas of the photo. Enhance the picture with any available tool, if required.

Findings from questions:

There are several crystals or crystal-like objects. How many colors and shapes of crystals are there?

Can you see the colored lenses? Where are they?

Look in detail at the image reflected on the spheres. Can you see the blurred image of the carriage house?

Can you see the reflection of the photographer, the camera or any tripod on the spheres? What does your answer tell you?

Finally, some photos are available in many books, and many other images can be found on the Internet to download and analyze. Scan or download some of these pictures at high resolution and perform your analyses.

CONCLUSION

After noting down your findings and answers to the questions think about what the detailed findings mean for yourself and humanity. Share and discuss your results and conclusions with others. Go over the exercises with them and compare their responses with your own. We suggest sharing your findings with a questioning, not an invasive mind.

Chapter 12

Experiment 10: Making a stereoscope for viewing 3D WCUFO pictures

INTRODUCTION

This simple, practical experiment helps show how 3D images work, how to see the WCUFO in 3D, and finally how to confirm details in the WCUFO sphere reflections.

It is fascinating and revealing to see 3D stereoscopic images of the WCUFO and its sphere reflections. Initially, nobody observing these WCUFO photos realized how important and revealing the sphere reflection details would be. Although not polished to a sheen each sphere still produces revealing reflected images of the ship's surroundings. They are "eyes" showing what they see.

We humans and most animals observe in 3D because we have two eyes. The composite image from each of them creates impulses in our brain enabling us to detect how near or far objects are. Covering one of our eyes for a few minutes while continuing with our normal life, soon results in difficulty determining the location and distances of different objects around us.

When looking at everyday photos, we cannot be sure how far away from the camera the objects were because the camera's single lens functions as a single eye. So in this experiment, we show how to build a stereoscope, take 3D photos, and how to use your homemade device to see and confirm WCUFO sphere reflection details in 3D.

How does this stereoscope work? As mentioned, the "magic" of seeing a picture in 3D relates to sending different images to each of our eyes. Figure 69 shows a simplified rendition of some trees from two separate locations. The two cameras, side by side each correspond to a single eye of a pair of eyes. Each one sees a slightly different image from its respective viewpoint. The two images are shown below in figure 69. Note their similarity, but the slightly different location of each tree. Looking at these pictures through a stereoscope produces a 3D image in our brain.

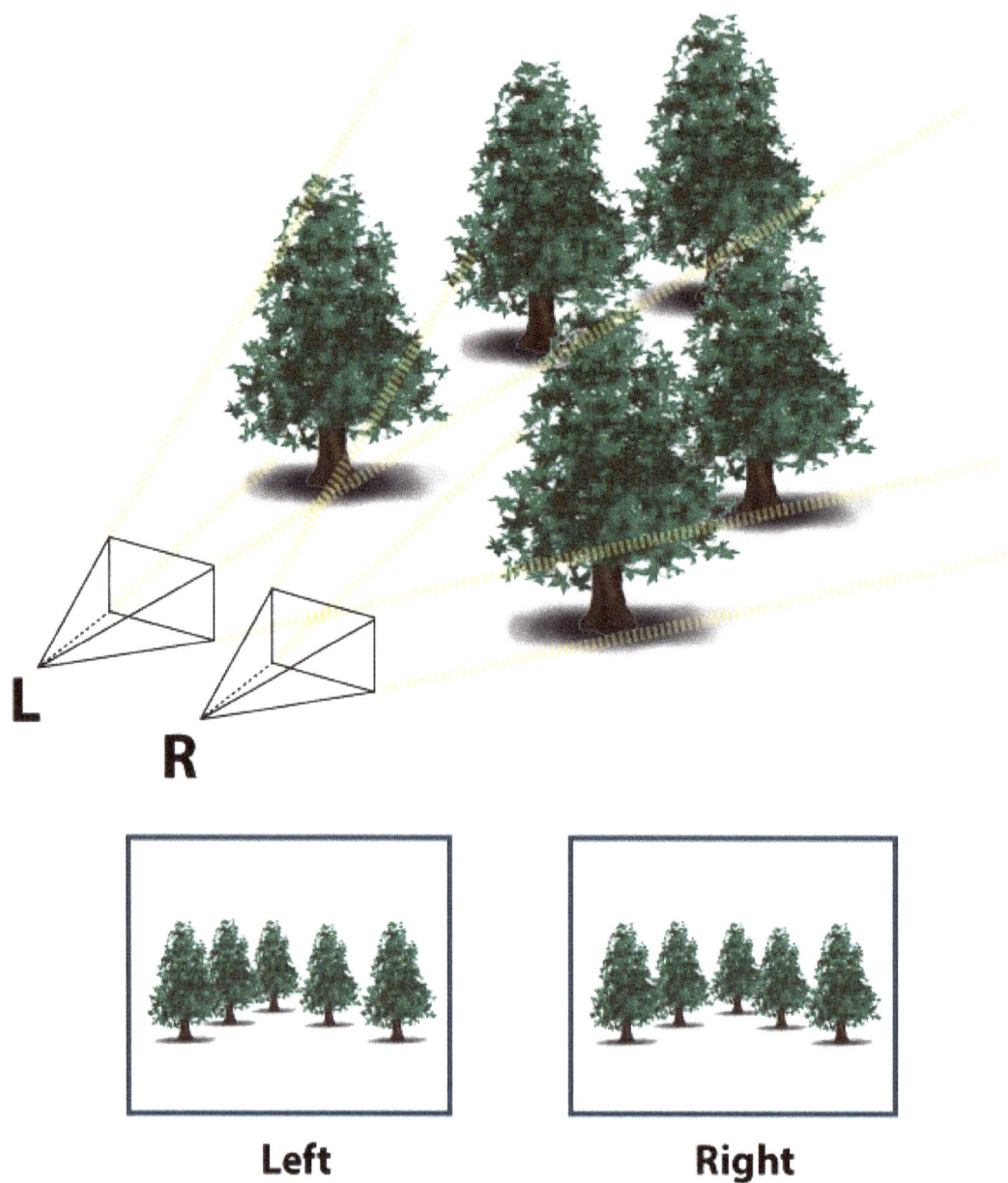

Figure 69- Simplified rendition of a forest. Two camera viewpoints **L** and **R** side by side correspond to the observation of a pair of eyes. Each sees a different image. Combined they give a 3D perception in the brain.

Some people have the ability to see these images in 3D without the aid of an instrument. They can look with the eyes in a lateral view instead of the usual view of converging to a point. The left eye must see the left figure and the right eye the right figure. It is like looking at a very distant object while the eyes focus on a nearby image. If like most, you find this difficult to do, the stereoscope is for you. Here is how to make and use one.

MATERIALS

All you need to make your stereoscope is: two identical magnifying lenses like the ones in either Figure 70 or 73, some plastic glue, a sheet of stiff card, a card cutter, a ruler, and some pictures to view.

METHOD

Stereoscope construction

Here are two methods for constructing a simple homemade stereoscope.

First method

Figure 70- Two identical magnifying lenses.

Avail yourself of two identical magnifying lenses somewhat like the ones in Figure 70. A magnifying lens has a set focus distance, so first we must find this. Look at the page of a book through your lens and when you see a very clear magnified image measure the distance from the lens to the page. Take note of this: it is your "focus distance."

Some magnifying lenses come with a frame on the base, conveniently elevated at the focus distance.

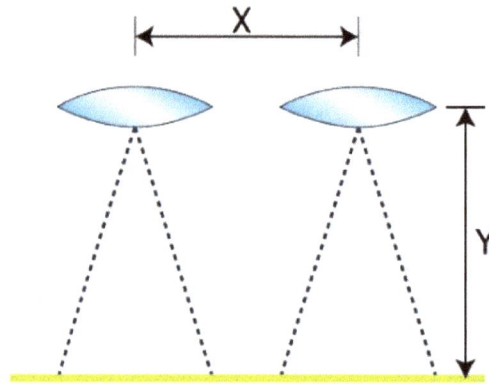

Figure 71- Stereoscope design with lenses separated by distance **X** (60 mm to 70 mm). **Y** is the focus distance.

To build a stereoscope both lenses must be located at the same height (focus distance **Y** in Figure 71), and separated by a distance similar to that of the eyes (**X** in Figure 71), that is 60 mm to 70 mm (2.4 to almost 3 inches).

Figure 72- Model stereoscope. The bases with frames removed and both pieces glued together.

Figure 72 shows a stereoscope made with two magnifying lenses. Eliminate the bases and frames and glue the pieces together keeping a distance of 70 mm between the lens centers.

Second method

Another way to make a stereoscope is using cardboard to make a base for two magnifying lenses as shown in Figure 73.

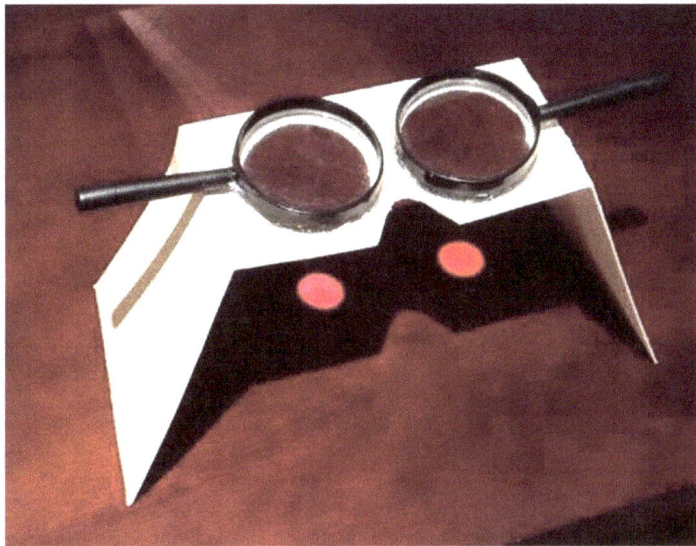

Figure 73- An alternative model stereoscope: Two magnifying lenses atop a cardboard base, their centers separated 70 mm at a height equal to their focus distance.

The magnifying lens should not be too big or too small. The diameter must be less than 5 centimeters (2 inches). If too big, their separation is too far for practical use. The focus distance must be between 50 mm to 100 mm (2 to 4 inches).

To verify the stereoscope works well, test it over Figure 69. The left and right lenses must be directly above the corresponding frames. You may need to increase or decrease the separation of the left and right images. To do this take a color copy of Figure 69. Cut and separate the two images. Then place them below the stereoscope and adjust their separation distance until the pictures overlay perfectly.

Taking stereo pictures

Now try making interesting stereo photos and viewing them in your stereoscope. Go outside to a static scene; it could be in nature or before some interesting architecture of a town or village. Take one photo, then take one step to one side and take another photo. Print both in small format, and put them below your stereoscope. Vary the separation of the photos until a unique 3D image appears.

If taking two pictures of a group of people, ask them not to move and to "freeze" any smiles while you take the two photos. Otherwise, the stereo image gets ruined.

When taking two pictures of a distant object, like a mountain, take one photo, then move several meters to one side and take the other.

Recommendations:

- For each pair of your pictures, select an object centering it in each shot. Centering the object frames both pictures in the same area.

- The camera separation distance between each photo should be 1/10 to 1/50 of the distance to the object photographed. For example, taking a picture of somebody at 5 meters away, the distance between the two shots would be between 10 cm to 50 cm.

- Ensure, of course, that the picture taken on the left is on the left under the stereoscope, and the right picture is on the right. With mismatching pictures, the 3D effect does not work properly.

WCUFO stereo pictures

Figure 74- The WCUFO in 3D.

Figure 74 shows a good 3D stereoscope view of the 3.5-meter WCUFO. Meier says he was on a bigger WCUFO close to it, both at treetop level. Since the Secondary WCUFO on which Meier sat taking the photos was moving, each image is from a slightly different location. We made this pair from Photos #838 (right) and #839 (left). Some details of distant mountains in one photo and a nearby branch in the other are removed to present an open 3D view of the WCUFO.

Each WCUFO sphere reflects an image of its surroundings, and since each sphere is in a slightly different location in space, they each have a unique image. Pairs of reflecting spheres also function as eyes observing what is around, albeit with blurred vision. Combining these images produces stereo views.

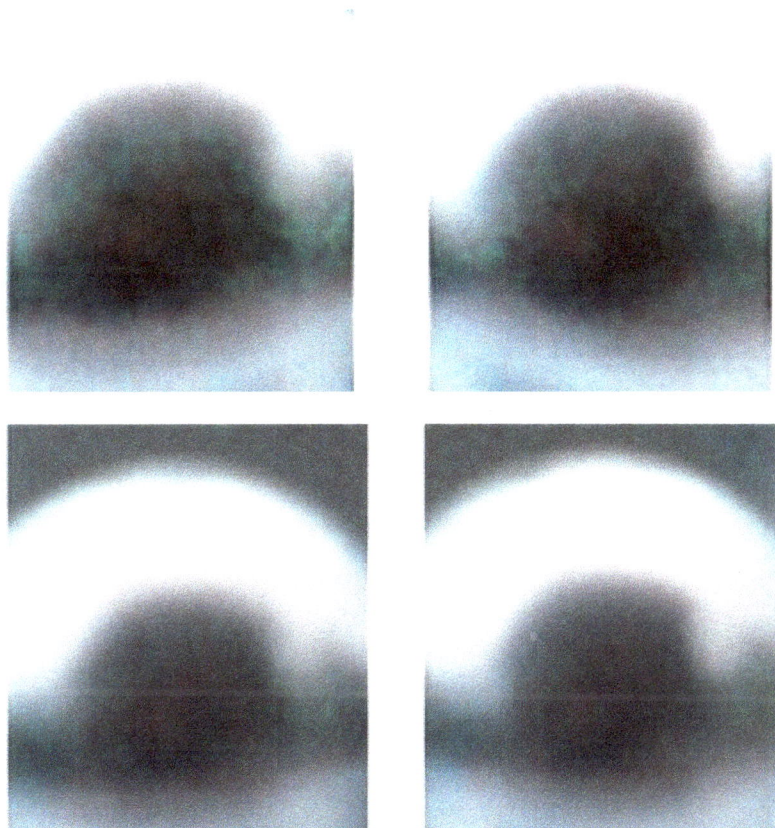

Figure 75- Blurry carriage house reflections on WCUFO spheres in Photo #808.

Figure 75 has zoomed reflection images from Photo #808 WCUFO spheres. Looking at them in our stereoscope, we notice a little central mass which must be Meier. At the far right of the carriage house, plants or trees are discernable that are farther away than the wall.

If Meier had used a scale model for his WCUFO, a sizable reflection of the photographer (or tripod) would be visible in these pictures, whereas none is visible in Figure 75, only a small brown figure in the center of the sphere images, which must be Meier. Its small size proves the WCUFO to be a significant and real WCUFO, not a small model.

More stereo pictures appear for you to view in the following pages. Some show a forest around the WCUFO, with one tree closer to the camera. Pairs of photos taken from Photos #800, #799, #839, and #834 (Figure 76) give combinations of different WCUFO sphere images for 3D viewing.

The four Figures 77 – 80 provide the photo pairs. Photo #800 pairs are in Figure 77, Photo #799 pairs in Figure 78, Photo #839 in Figure 79, and Photo #834 in Figure 80.

Figure 76- Photos #800, #799, #839, and #834 used to create the stereo pairs in Figures 77-80.

Photo #800

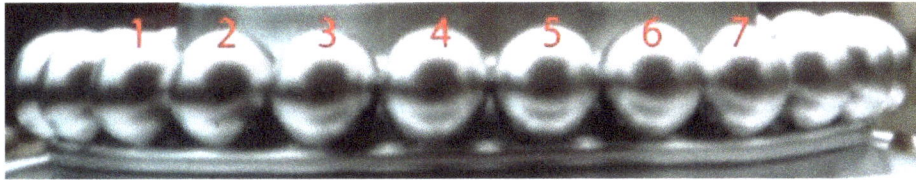

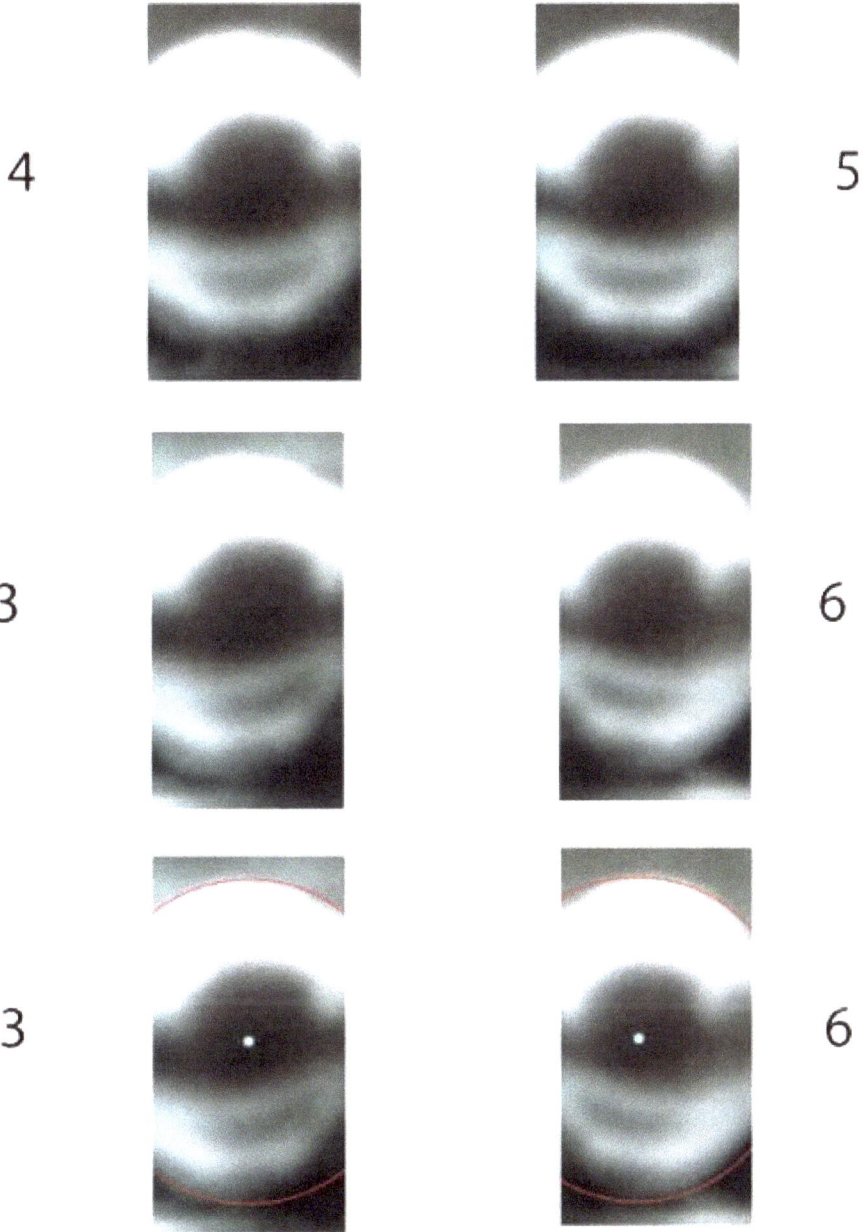

Figure 77- Stereo pairs, #800. Blue dots at the center of the spheres on the bottom row are the camera's location observed by the camera.

Photo #799

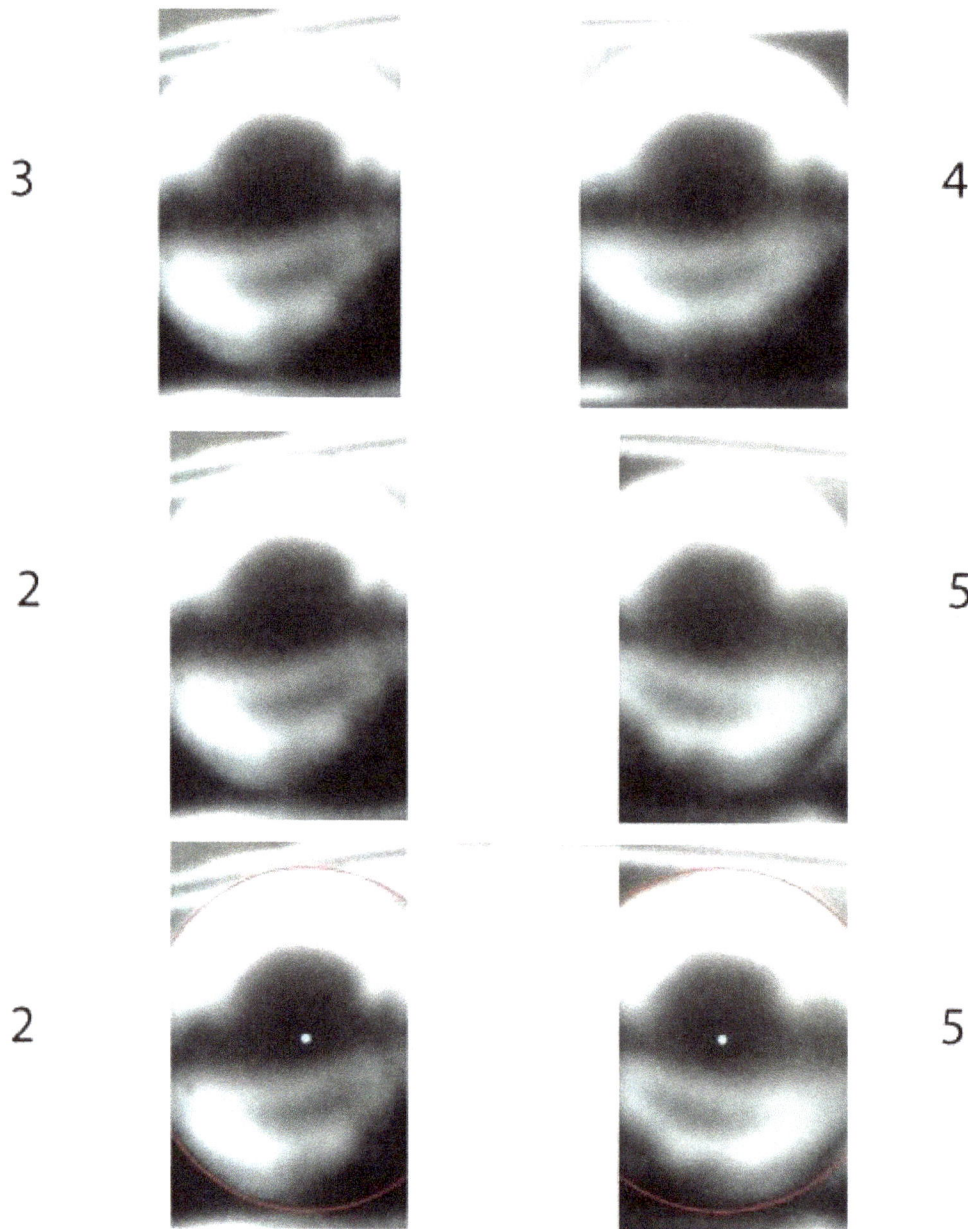

Figure 78- Stereo pairs #799. Blue dots at the center of the spheres on the bottom row are the camera's location.

Photo #839

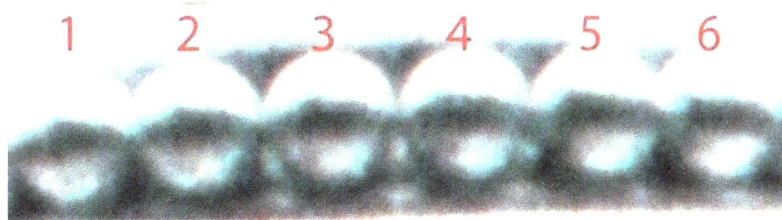

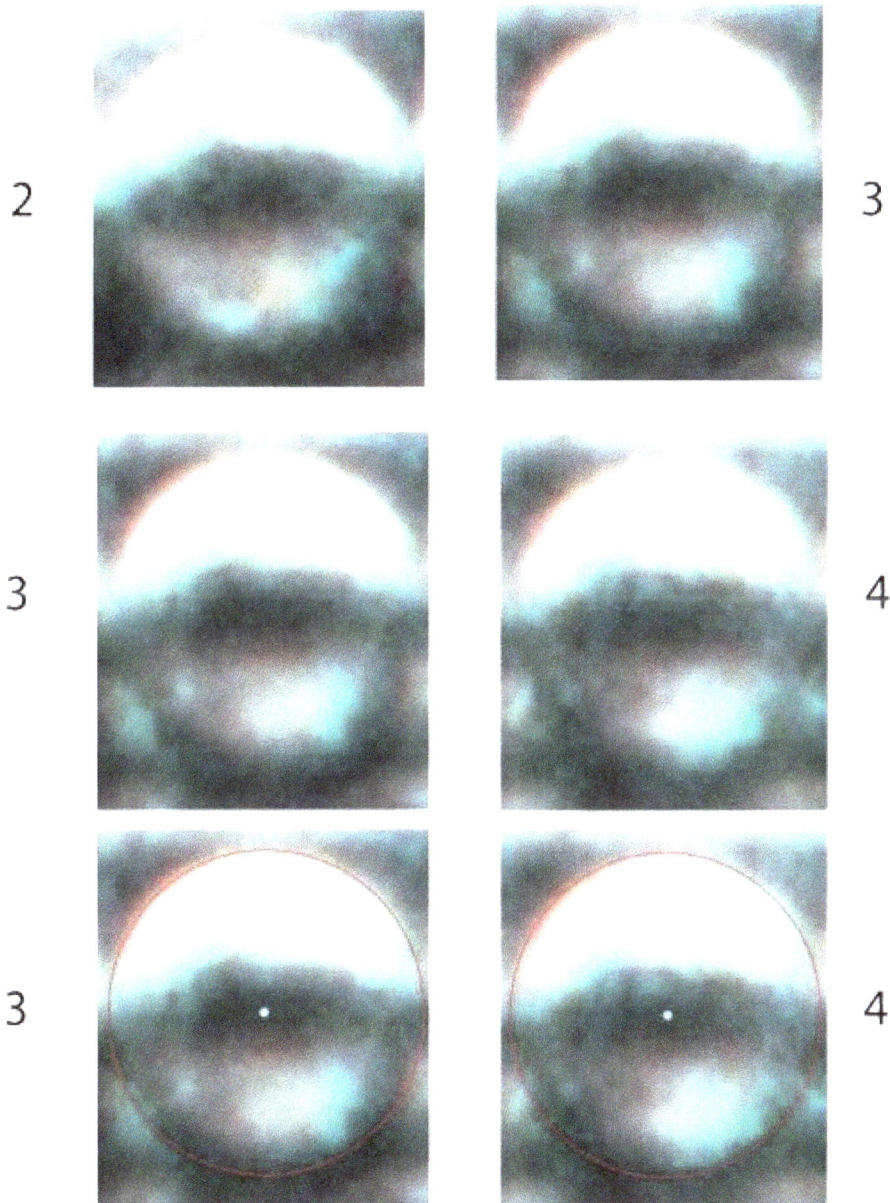

Figure 79- Stereo pairs #839. Blue dots at the center of the spheres in the bottom row show the camera's location.

Photo #834

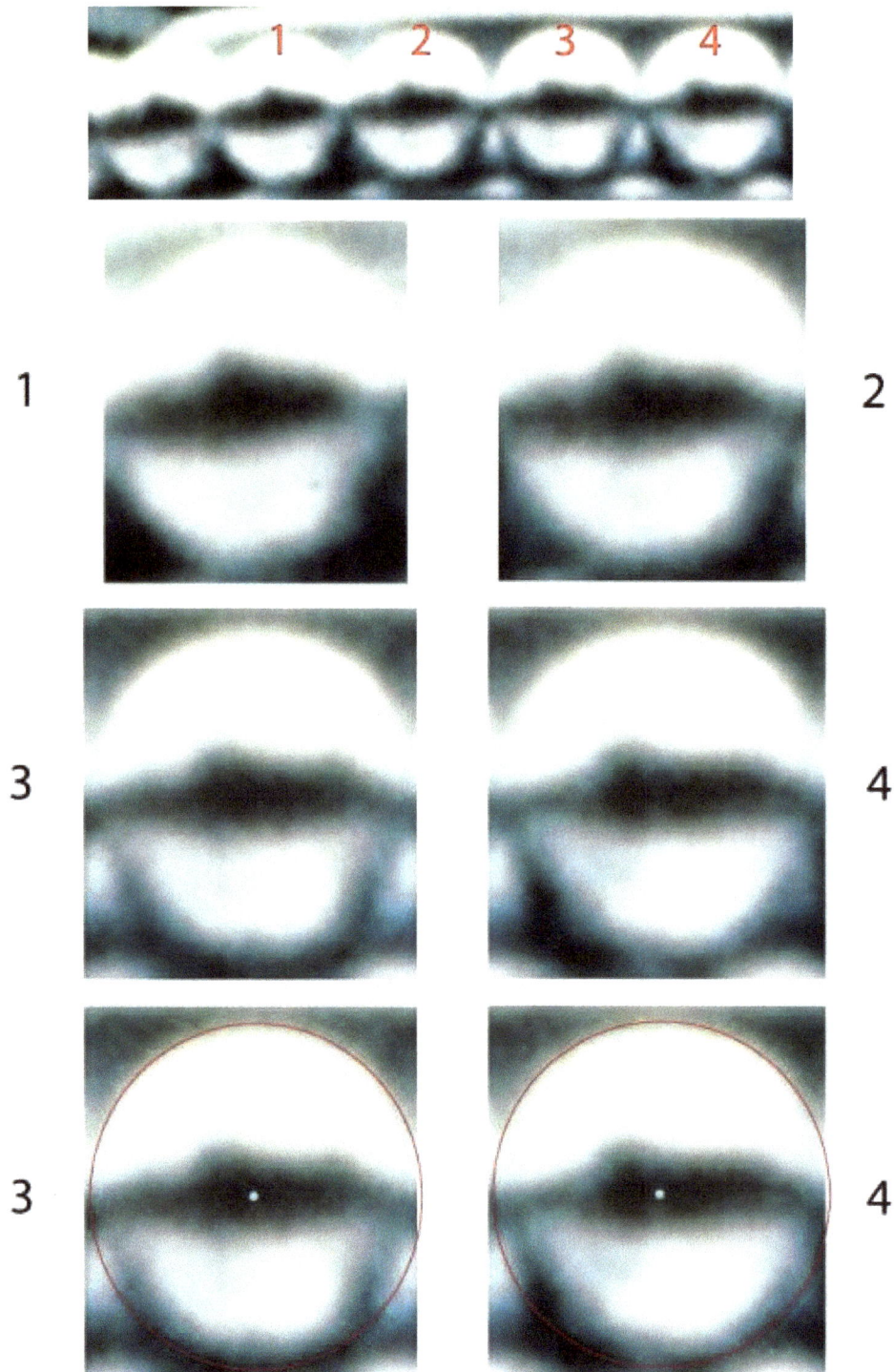

Figure 80- Stereo pairs, #834. Central blue dots on the bottom spheres show the camera's location. A nearby treetop is in the middle just left of center.

Making personal stereo photos of the WCUFO

Anyone can create such stereo pairs. All you have to do is find good WCUFO illustrations in a book, take close up photos, increase the contrast and brightness, and construct the pair. One good pair is possible from Photo #834 from a quality reproduction in the book *Through Space and Time,* page 25 (Meier 2004), or *Zeugenbuch* page 271 (Meier FIGU 2001). Also, several WCUFO photos are available in *Photo-Inventarium* pages 103 to 125 (Meier FIGU 2014). Alternatively, you might enjoy constructing additional pairs from available digital copies.

Finally, in most WCUFO stereo photo pairs, trees or the carriage house profile are seen. In spherical mirrors, reflected images look small (or far away), so they also look small in the WCUFO spheres. To an observer inside the WCUFO, these objects would not appear so small, or so far away. Somebody viewing them from the WCUFO would see how big and how high the trees are. To appreciate what they saw refer to Chapter 10 Experiment 8, and the method of finding the elevation or height of nearby reflected objects like a tree.

Chapter 13

Experiment 11: Height changing capability of a WCUFO

INTRODUCTION

We have seen that WCUFOs photographed by Billy Meier come in two different sizes: 3.5 m and 7 m. All WCUFO versions look about the same; however, in one case a WCUFO central core appears to have extended a little upwards.

So in this Experiment, based on an analytical comparison between a Meier nighttime photo and a daytime photo, we determine whether a WCUFO central core has risen in this picture, and if so the height of its upward extension.

Figure 81- Photo #873 WCUFO on the night of August 5, 1981.

The WCUFO Photo #873 in Figure 81 taken on the night of August 5, 1981, gives the impression that its central core is vertically extended a bit compared with the WCUFO in other photos.

Another WCUFO seen in Photo #999 shot at night a few days before on August 2, 1981 (Figure 82), gives the impression that its core was not elongated upwards like the WCUFO in Photo #873.

Figure 82- Photo #999 WCUFO on the night of August 2, 1981.

Is the WCUFO core in Photo #873 extended, or is this an optical illusion caused by perspective? Alternatively, is the appearance due to the bright and dark bands on the WCUFO core? In this experiment, you answer this question by making a geometrically based drawing and applying analytical methods.

MATERIALS

Photo #873 (Figure 81 WCUFO at night), Photo #800 (Figure 84 daytime WCUFO), and a ruler to make measurements.

METHOD

First, we check the distance between two different platforms: the upper, with its mid-level spheres, indicated by the red ellipse, and the lower platform with its largest spheres indicated by the green ellipse. We define this lower level outer edge at just outside the row of tiny red crystals. See Figure 83.

Figure 83- Vertical separation distance between the mid-tier and lower tier.

Use the lower tier sphere diameter as the unit of measurement, so every measurement made is expressed in these sphere diameters. Measure the diameter of these spheres marked with pink dashed circles. Use the image on this page, or an electronic copy of this photo zoomed on your computer screen.

Segment E-D in Figure 83 is not in front view but tilted. We calculate at around 17 degrees. The error induced by this inclination is around 6%, which is not big but is important to consider. So while remembering it, for our calculations we ignore this WCUFO tilt for now.

We calculate the vertical separation between the two platforms, in sphere diameter units by measuring distance E-D, and dividing it by the sphere diameter:

$$\text{Separation of Platforms} = \text{Segment E-D} / \text{Sphere size.}$$

Therefore the **platform separation is 2.03** sphere diameters.

We use the major axis of each ellipse to locate each platform. Due to the previously noted error of 6%, this value could be increased by 6%. In our calculation, we obtained a greater value of 2.15.

Now we compare this platform separation with the separation in Photo #800 in Figure 84 where the WCUFO is viewed almost on edge or at eye level. The blue line segment delineating the separation between the platforms here is in frontal view reducing the considered 6% error in measurement in Figure 83 to near zero.

Performing the same calculation for Photo #800:

$$\text{Separation of Platforms} = \text{Blue Segment} / \text{Sphere size}$$

Therefore the **platform separation is 1.91** sphere diameters.

You might find like we did that the separations are not the same. The WCUFO in Photo #873, photographed at night, has a bigger separation. We estimate the difference is around 0.25 sphere diameters. In other words:

The WCUFO in Photo #873 extended its central core upwards by a distance equal to one-quarter of a sphere diameter.

What do *you* conclude? Does the bigger 7 m diameter WCUFO have a taller central core than the 3.5 m WCUFO? Do the same analysis on other WCUFO photos, like #999 in Figure 82. Does Photo #999 display the same extended central core? Was this extension only on Photo #873?

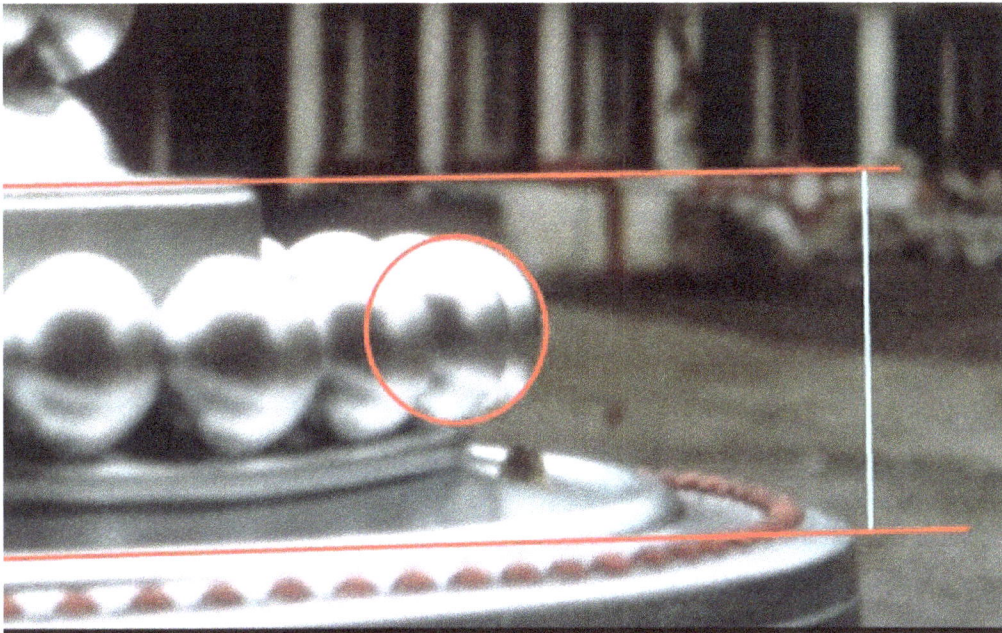

Figure 84- Vertical separation between the two platforms. Photo #800.

Another question to consider: Is it feasible to endorse the idea of a small UFO model having a capability – unmentioned and unnoticed for over 30 years – of extending its central core? It is not impossible, but it introduces yet another strong problem for the already disproven 55 cm UFO model theory.

Final words and post-experiment note

Here concludes our guide through the eleven experiments for this mysterious UFO. We encourage you firmly to perform other tests for yourself. Having performed the experiments as a researcher or budding scientist now try additional tests to share with friends, students, and other investigators.

The enquiring scientific mind that requires answers through tests and experiments needs no special qualification. What counts is the quality of one's research, testing and experimentation. Remember, anybody carefully analyzing this evidence, and testing with simple tools can be a research scientist.

For a researcher, having an open yet skeptical mind to consider the findings of some UFO investigators regarding this WCUFO is a good start; but it is far better to perform personal research, and especially tests and experiments that after assiduous work bring personally verified sound conclusions.

In closing, we suggest employing patience, care, concentration and understanding as you share your findings with others.

Finally, for your interest we share part of Contact Report #442, of February 10, 2007, in which Ptaah eventually gave some measurements of the WCUFO hovering above Meier's parking lot (Frehner email, German trans. Devine):

Billy

The same goes for my side. - Some time ago, Professor James Deardorff, from the United States, wanted to know some things regarding the cake ship that was landed in our car park between the shed and the house. The history - in regard to the striking similarity of the ship's hull with a barrel cover shape, that is to say, with a drum lid, and the emergence of those storage drums which we discovered on our property - is listed in the 254th contact report. Because of that, Christian Frehner had to take various photos to specify how large the ship was and the distance between me with the camera and the cake ship. For my part, I know that the diameter was approximately 3.50 meters, which I also told Christian. But at the same time, can you now also provide for me the exact measure, as well as how many people can be accommodated in this small ship?

Ptaah

3. This is no secret, and I know the precise data:

4. The lowermost outermost diameter, with the flat edge, was 3.52 meters, however, the outermost upper outer rim

Billy

Das gilt auch meinerseits. – Vor geraumer Zeit wollte Professor James Deardorff aus den USA einiges wissen bezüglich des Tortenschiffes, das auf unserem Parkplatz zwischen der Remise und dem Wohnhaus gelandet wurde. Die Geschichte in bezug auf die frappierende Gleichheit des Rumpfes des Tortenschiffes mit einer Fassabdeckform resp. mit einem Fassdeckel und dessen Entstehung von sich in unserem Besitz befindenden Lagerfässern ist im 254. Kontaktbericht aufgeführt. Christian Frehner musste deswegen verschiedene Photos machen und angeben, wie gross das Schiff und die Distanz zwischen mir mit dem Photoapparat und dem Tortenschiff war. Meinerseits weiss ich, dass der Durchmesser rund 3,50 Meter beträgt, was ich Christian auch sagte. Kannst du mir dazu nun aber noch das genaue Massangeben, aber auch wie viele Personen in diesem kleinen Schiff Platz finden?

Ptaah

3. Das ist kein Geheimnis, und die Daten sind mir genau bekannt:

4. Der unterste äusserste Durchmesser mit dem flachen Rand betrug 3 Meter und 52 Zentimeter, der äusserste obere Aussenranddurchmesser jedoch 3 Meter

diameter was 3.20 meters.

und 20 Zentimeter.

5. The entire outer edge structure, on which the swinging-wave accumulators were attached, was 37.6 centimeters, while also with this, the measurement of the bottom ring to the level of the swinging-wave accumulators was 32 centimeters.

5. Der gesamte äussere Randaufbau, auf dem die Schwingungskumulatoren angebracht waren, betrug 37,6 Zentimeter, während das Mass vom unteren Randring bis zur Ebene der Schwingungskumulatoren 32 Zentimeter betrug, wie auch bei diesen.

6. And room in the flying device was designed for a person in a sitting position, but in an emergency three persons could find room in a crowded manner.

6. Und ausgelegt war der Platz im Fluggerät für eine Person in sitzender Stellung, wobei aber notfalls drei Personen in gedrängter Weise Platz finden konnten.

7. These types of flying devices were not suitable for the Earth's atmosphere, for which reason, after a brief time of service, they were withdrawn from terrestrial space.

7. Diese Art Fluggeräte haben sich für die irdische Atmosphäre jedoch nicht geeignet, weshalb sie schon nach kurzer Einsatzzeit wieder aus dem irdischen Raum abgezogen wurden.

Billy

Billy

That was not known to me. With the swinging-wave accumulators, the silvery spheres are probably meant, I suppose, or?

Das war mir nicht bekannt. Mit den Schwingungskumulatoren sind wohl die silbrigen Kugeln gemeint, nehme ich an, oder?

Ptaah

Ptaah

8. That is right.

8. Das ist richtig.

9. Their diameter was 32 centimeters, as I already said.

9. Deren Durchmesser betrug 32 Zentimeter, wie ich bereits sagte.

Billy

Billy

Thanks for the data, which Christian can probably use.

Danke für die Daten, die Christian vielleicht gebrauchen kann.

Here we see specific measurements of this WCUFO. It also states that the WCUFO can hold one pilot or three in a crowded manner in an emergency. Perhaps the WCUFOs withdrew due to corrosion caused by our atmosphere. Perhaps with a cover, like other beamships, they might be more suitable and last longer. Interestingly Ptaah refers to the spheres as "swinging-wave accumulators." We understand these as something akin to frequency or resonance accumulators.

Annexes

Annex A
WCUFO proportions

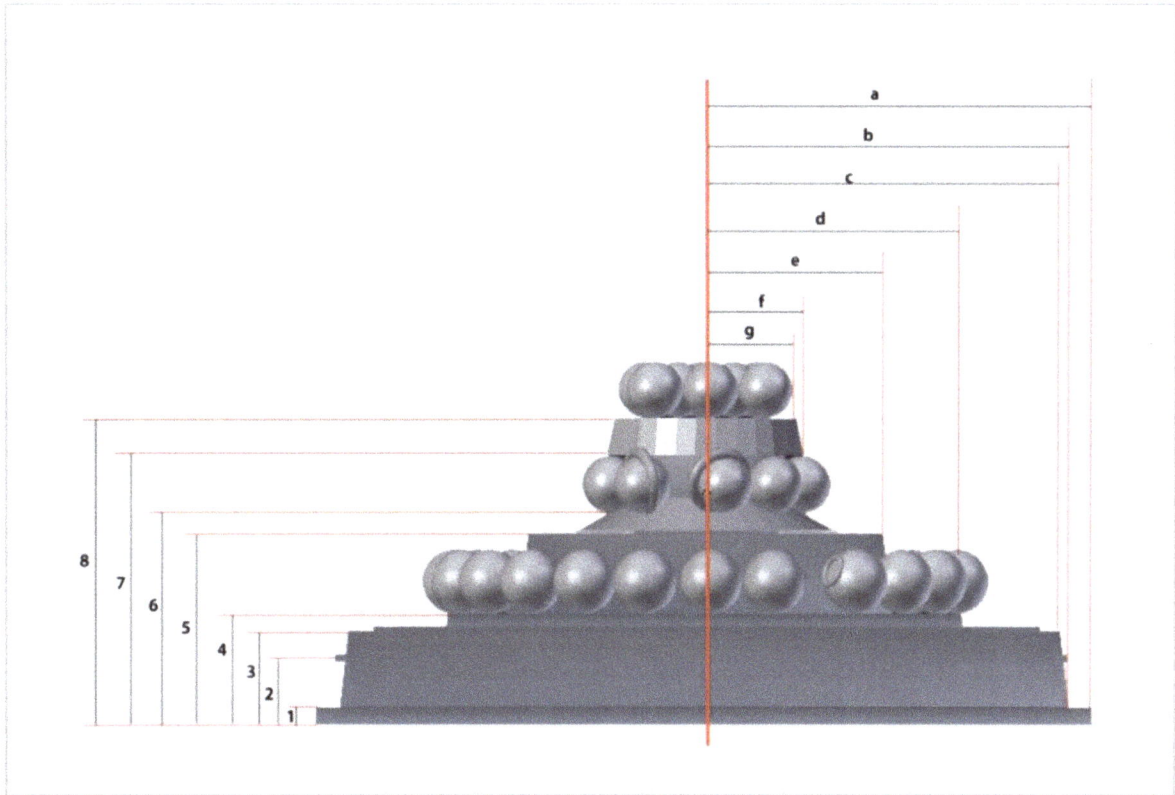

Figure A1- Vertical and horizontal distances.

The measurements for WCUFO proportions (see Figure A1) come from Photos #800 and #834 for the 3.5 m diameter WCUFO. Distances marked as 0, 1, 2, 3, through 8 were not calculated for #800 since they were not all visible. Measurements for the 7 m diameter WCUFO come from Photo #999.

Dia 1 (see Table A1) is the diameter of the lower circle of spheres, Dia 2 of the middle circle of spheres, and Dia 3 is the diameter of the upper circle of spheres.

| Feature | WCUFO 3.5 m | | | | | | WCUFO 7m | | |
| | #800 | | #804 | | | | #999 | | |
	Dim 1	Norm 1	Dim 2	Norm 2	Avg.	R	Dim 1	Norm 1	R
a	206.0	6.65	244.5	6.35	**6.50**		177.0	**6.81**	
b	196.0	6.32	235.0	6.10	6.21		170.5	6.56	
c	191.0	6.16	231.5	6.01	6.09		166.5	6.40	
d	121.5	3.92	155.5	4.04	**3.98**	0.61	108.0	**4.15**	0.61
e	86.5	2.79	111.0	2.88	2.84		75.0	2.88	
f	51.0	1.65	64.0	1.66	**1.65**	0.25	46.5	**1.79**	0.26
g	46.0	1.48	57.0	1.48	1.48		42.0	1.62	
0			0.0	0.00	0.00		0.0	0.00	
1			7.5	0.19	0.19		5.0	0.19	
2			37.0	0.96	0.96		23.5	0.90	
3			50.0	1.30	1.30		35.0	1.35	
4			64.0	1.66	1.66		49.0	1.88	
5			113.0	2.94	2.94		85.0	3.27	
6			136.0	3.53	3.53		101.0	3.88	
7			172.0	4.47	4.47		128.5	4.94	
8			197.0	5.12	**5.12**	0.79	146.0	**5.62**	0.82
Dia 1	31.0	1.00	38.5	1.00	1.00		26.0	1.00	
Dia 2	29.0	0.94	36.0	0.94	0.94		24.0	0.92	
Dia 3	27.0	0.87	35.5	0.92	0.90		23.0	0.88	

Table A1- WCUFO measurements taken from several photos.

Dim. 1 and Dim. 2 are the two measurements taken from two 3.5 m WCUFO (Photos #800 and #834) in millimeters calculated on a flat computer screen. Norm 1 and Norm 2 are the normalized values of all measurements, using the diameter of the lower group of distinct spheres as the unit; this diameter is the basic normalizing unit (value = 1). The Avg. column shows the normalized value averaged for both 3.5 m WCUFOs analyzed. For example, the radius **a** of the 3.5 m WCUFO is 6.5 times the diameter of the lower spheres.

R values are ratios of horizontal and vertical proportions:

Rd = ratio of the distance of the lower-level spheres to the central axis, and the WCUFO radius. (**Rd** is **d/a** in Figure A1, 0.61 value)

Rf = ratio of middle-level spheres' distance to the central axis, and the WCUFO radius. (**Rf** is **f/a** in Figure A1. 0.255 average value)

R8 = vertical ratio: the vertical distance from the base to the upper platform with its spheres, and the WCUFO radius (**R8** is distance **8** divided by distance **a,** 0.805 average value).

When calculating the WCUFO diameter using the reflection method we ignore the WCUFO lower part and metrics here since it is not always visible and is unnecessary.

Rd (0.61) is used to derive the WCUFO diameter and distance to the camera by observing the nearest sphere reflections. (See Chapter 8)

It appears the 7 m WCUFO is somewhat taller and wider, in proportions than the 3.5 m WCUFO. Alternatively, maybe the lower spheres are a little smaller. We did not consider Photo #873 for these measurements because its central core presents a vertical extension (see Experiment 11). Photo #999 provided this calculation, and it does not show this upward extension.

Also of note is that spheres of different tiers have different diameters: 1.00 / 0.94 / 0.90 for the 3.5 m WCUFO; and 1.00 / 0.92 / 0.88 for the 7 m WCUFO. This variation is not solely the perspective effect of distant objects looking smaller than close ones. Change due to perspective should not be more than 3% in the WCUFO above the treetops (photo #834, Figure D4), and not the 8% photo #834 shows. The spheres are of different sizes; the lower tier spheres are bigger than those in the middle tier, and the spheres at the very top are the smallest.

In our opinion, making a little WCUFO model using Christmas balls with slightly different sizes of up to a mere 12% would have been very complicated if not impossible in 1980. Christmas tree balls of the late 1970s did not, as far as we know, come in such slight variations of size.

Annex B
Carriage house dimensions
and courtyard plan view

The carriage house

The WCUFO spheres reflect images of the carriage house northeast wall. Knowing the size and orientation of this carriage house is paramount for determining the size of the WCUFO.

Figure B1 shows the dimensions of the carriage house northeast wall, measured by Christian Frehner on Meier's property.

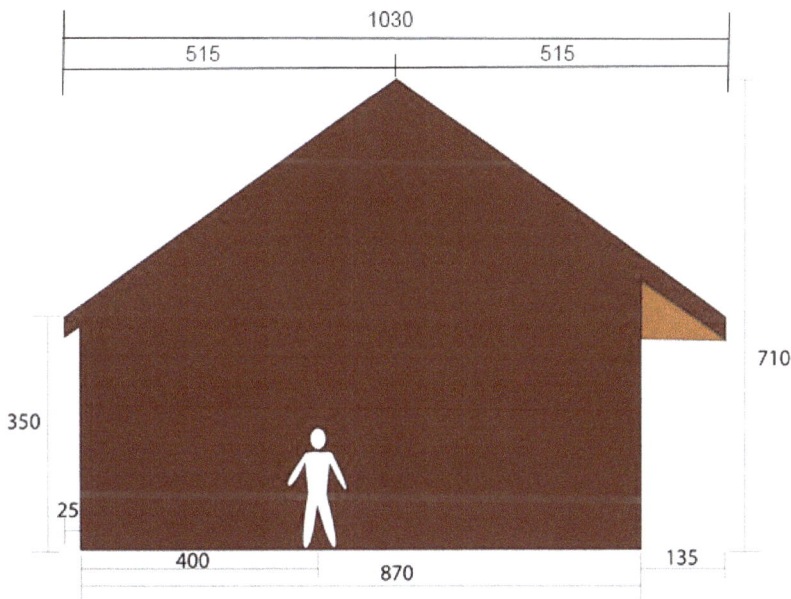

Figure B1- Dimensions of the carriage house northeast wall in centimeters.

The carriage house has experienced modifications over the years. Figure B2 is a picture taken circa 1977 some years before Meier took his WCUFO pictures. By this time, the southeast extension had already been completed.

Figure B2- Carriage house southeast extension circa 1977.

In 1980 when Meier took his WCUFO pictures the construction included this extension. Confirmation comes from a scene in the late 1970s Larry Savadov movie *Contact* created from investigations by Lee and Brit Elders, Wendelle Stevens and Tom Welch and now available online as *Contact from Pleiades Film*. At the end of the movie, in which Lee Elders and Wendelle Stevens talk about the then recent WCUFO photos, a scene shows the carriage house with its southeast extension (1:27:30~, 1:35:30~).

Figure B3- Carriage house scene showing southeast extension from *Contact From Pleiades* (at 1hr 35mins).

We see the courtyard between Meier's main house and the carriage house had no trees in the late 1970s. Currently, there are large, healthy trees in this area. Sometime after 1980 the carriage house had another extension, this time on its first-floor northwest side. Looking at recent (2015) Google Earth images, we see the new trees. Figure 19 shows a Google Earth aerial view of Meier's house in the upper right corner, and the carriage house in the bottom left corner, where again we see a part of the northwest extension roof.

Christian Frehner took the recent picture below (B4) showing the carriage house with the first-floor northwest extension (on the right).

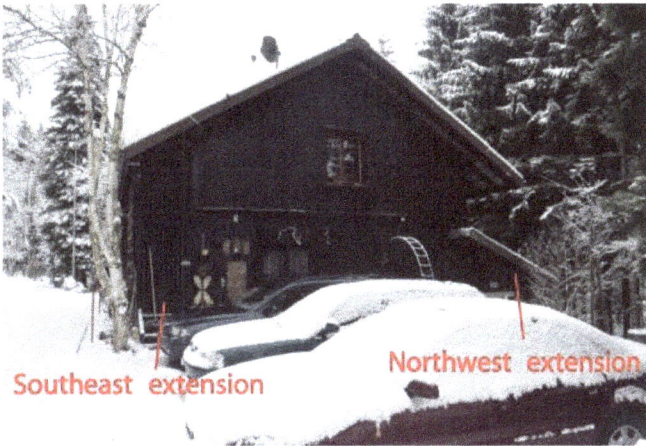

Figure B4- A 2013 carriage house picture showing the northwest extension.

Courtyard plan view

Below (Figure B5) gives a wide-angle view of Billy Meier's home on the right and the carriage house on the left in 1981.

Figure B5- Wide angle view of Meier's prime property in 1981.

With Google Earth images, available pictures, scenes from *Contact* movie and on-site measurements, we generated a plan view map of this area as it was in 1980. See Figure B6. On this map, Meier's house is at the top and the carriage house at the bottom. On the northeast wall of the carriage house, a red dot denotes where Meier was taking the WCUFO photos. See Chapter 5, Experiment 3 how we know Meier took his pictures from here and not farther away from the carriage house wall. The angles show the direction of the WCUFO projected to the main house; picture #800, blue, and #799, red.

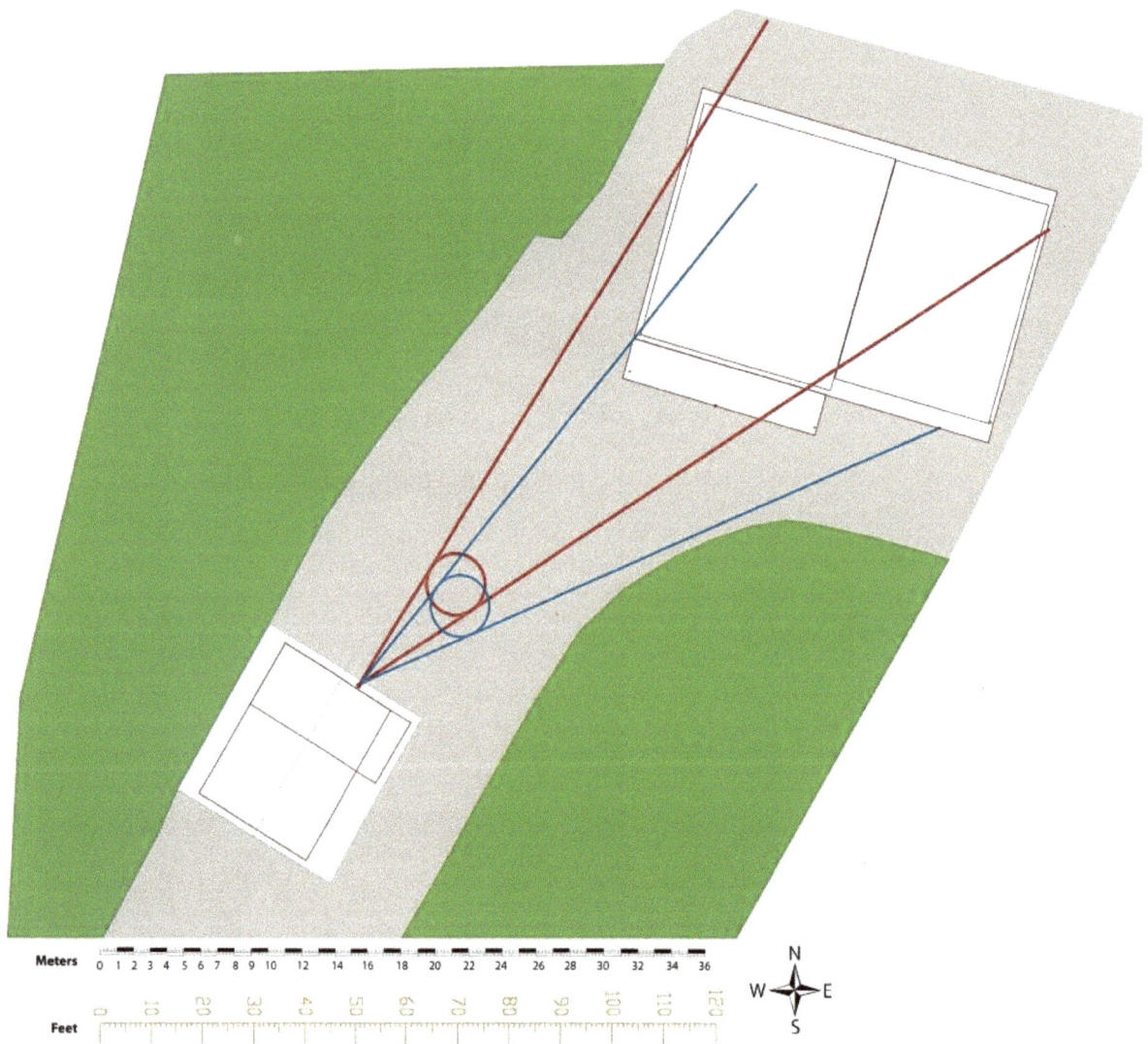

Figure B6- 1980 courtyard plan view generated with Google and other updated plans of Meier's residences.

The red and blue 3.5 m diameter circles are the location of this WCUFO (Photos #799 and #800) assuming it was 3.5 m in diameter.

The reflected image of the carriage house wall is very similar to the one in the analysis done by the computer program for making 3D models including reflections (Zahi and Lock *They Are Here*), and by using small spheres in a scale model of Meier's property. As we now know: the size of the sphere does not matter, since *the magnitude of the reflected image is directly proportional to the sphere's diameter* (sphere reflection **Rule c**).

In the computer program "Blender" we made a 3D model rendition of the WCUFO, the carriage house, and Meier's main house. (See Figure B7.) These helped us make the analysis of the WCUFO sphere reflections presented in *They are here*. In this book, we have not covered the use of a 3D modeling tool as another Experiment because it requires advanced skills. If, however, you are interested in experimenting with Blender, read *They are here* and contact the authors to receive a digital copy of the 3D models created.

Figure B7- "Blender" computer model of the carriage house and main residence.

Annex C
Test Sphere band separation

When using a reflecting Test Sphere as in Experiment 5, we stick two thin vertical strips of tape on it to match the width of the carriage house wall, or roof. Below is the formula that correlates the separation of the two vertical strips with the expected size of the carriage house wall reflection.

Front view:

Top view:

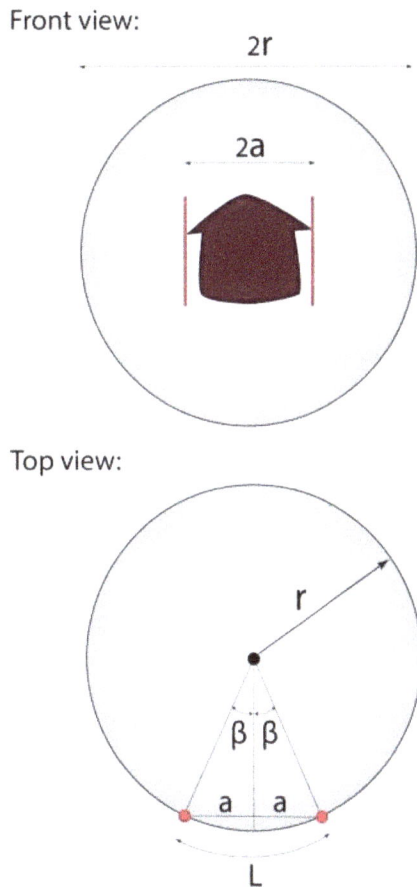

Figure C1- Separation of vertical reference lines.

Figure C1 (top) shows the two vertical lines separated a distance of twice the value of **a**. The radius of the sphere is **r**, so the diameter is **2r**.

As defined before, **Rch** is the relationship between **2a** and **2r**:

$$Rch = \frac{2a}{2r} = \frac{a}{r}$$

The size of the reflected carriage house image indicates how far or close the sphere is to the construction. So the size of the reflection helps establish the WCUFO size. The smaller **Rch** is, the farther away this WCUFO is and the bigger its size.

From this formula we can obtain:

$$a = r\,Rch$$

The curved distance between the two strips on the Test Sphere's surface is **L** (Figure C1 bottom). **L** is a bit bigger than **2a**. **L**'s value is a function of the sphere radius **r** and the angle **β**, as follows:

$$\frac{L}{2\beta} = \frac{P}{360}$$

Where **P** is the perimeter of the Test Sphere (equal to 2 π r). From this we obtain:

$$L = \frac{P\,\beta}{180}$$

Also:

$$Sin\,(\beta) = \frac{a}{r}$$

Moreover, **a/r** is also **Rch** as we found before, so:

$$Sin\,(\beta) = Rch \quad or \quad \beta = Sin^{-1}(Rch)$$

Therefore by substitution:

$$L = \frac{P}{180}\,Sin^{-1}(Rch)$$

This formula relates the separation **L** between the two white strips with the size of the sphere's perimeter **P**. The values of **Rch** can be measured from the original WCUFO photos taken by Meier or from reproductions. See Experiment 5 explaining how the Test Sphere is used to find the distance from the WCUFO to the carriage house (matching **L** with the carriage house roof width) and to establish how big this WCUFO is.

Annex D
Large format WCUFO photos

Large format WCUFO photos referenced in this book (see *Photo-Inventarium* book for more).

Figure D1- WCUFO Photo #799 taken by Meier on October 22, 1980, at 11:23 am.

Figure D2- Top: WCUFO Photo #800 of October 22, 1980, at 11:23 am. Bottom: WCUFO Photo #808 of October 22, 1980, at 11:24 am.

Figure D3- Top: WCUFO Photo #841 of April 3, 1981, at 1:20 pm. Bottom: WCUFO
Photo #844 of April 3, 1981, at 1:35 pm casts a huge shadow over a tree.

Figure D4- Photo #834. April 3, 1981, at 1:10 pm.

Works Cited

Deardorff, James. *The Wedding-Cake UFOs.* http://www.tiresearch.info/Wecake.htm. April 2013.

FOM. *The Future of Mankind - A Billy Meier Wiki - Why the WCUFO?* http://www.futureofmankind.co.uk/Billy_Meier/Why_the_WCUFO%3F. N.p., n.d. Web. 17 Mar. 2016.

-------. *The Future of Mankind - A Billy Meier Wiki - Why the WCUFO?* "The Pleiadian/Plejaren Contact Reports." Contact Report 254. http://www.futureofmankind.co.uk/Billy_Meier/Contact_Report_254. 2015, Oct. 2017.

Frehner, Christian. "Contact Report 442" part. Email. Trans., Devine. 13 Sep. 2016.

Meier, "Billy" Edward Albert. *Contact Reports Vol. 3.* (*Plejadisch-plejarische Kontakberichte, Gespräche, Block 3*). 375 – 388. "Plejaren Contact Report 123." 4 June 1979.

-------. *Through Space and Time,* p25, 109. Tulsa OK: Steelmark LLC. 1 Jan 2004.

-------. *Photobuch. Wassermannzeit-Verlag,* p 114. Switzerland. 2001.

-------. *Photo-Inventarium,* p102, 103-125. FIGU. Augsburg: Wassermannzeit-Verlag, 2014.

-------. *Zeugenbuch.* p 271. Schmidrüti, Switzerland: FIGU, 2001.

Plumridge, Jo. "What is Pincushion Distortion?" https://www.lifewire.com/what-is-pincushion-distortion-493732. 9 March 2016.

Savadove, Larry. *Contact.* "Billy" Meier documentary film. BGR Entertainment Corp. Prod. Savadove Young Films. Assoc. The Phoenix Film Group. Investigators Lee and Brit Elders, Wendelle Stevens and Tomas Welch. 1978. Aka *Contact from Pleiades Film.* 1:35:30~. https://www.youtube.com/watch?v=8dhutxr7W4w. 14 Jan. 2014.

Stack Exchange. *Photography.* Post 29. http://photo.stackexchange.com/questions/12434/how-do-i-calculate-the-distance-of-an-object-in-a-photo. 24 March 2016.

Stevens, Wendelle. *Contact – 'Billy' Eduard A. Meier Documentary by Wendelle Stevens (1978).* YouTube. https://www.youtube.com/watch?v=vzG6B4uEuwA. 23 Feb 2013. 1:27:30 ~.

Zahi, Rhal. *Analysis of the Wedding Cake UFO: investigations of the WCUFO pictures taken by Billy Meier.* PDF file 4.0MB. March 2013. http://www.tjresearch.info/Zahi_WCUFO%20Investigation.pdf. See also: http://www.futureofmankind.co.uk/Billy_Meier/Analysis_of_the_Wedding_Cake_UFO. 6 Oct. 2013.

------- and Christopher Lock. *They are Here: Compelling evidence of ET presence on Earth.* Planned publ. 2017.

Glossary of graphic symbols

These final two pages contain the graphic symbols and their meanings in the text. **Basic skills** symbols colored blue below indicate skills that assist in performing the experiments or analyses, and the yellow **Key symbols** on the next page are the three key **Sphere Reflection Rules** or principles outlined in this book.

Basic skills

Very basic photography abilities: take simple photos with a basic zoom camera or smartphone and upload them to a computer.

Very basic computing skills: perform simple computer image processing.

Curiosity, exacting observation and critical thinking. Anyone can easily follow this section by exercising critical observation and thinking.

Simple model-making manual tasks: cut and glue cardboard parts to make simple scale models.

Knowledge of junior high school math and trigonometry: at times conduct simple math calculations or use basic trigonometry.

Drawing basic scaled plans: make sketches and plans keeping scales and proportions accurate to scale.

Knowledge of some photography: understand depth of field, exposures, lenses, camera geometry, and image processing.

Sphere Reflection Rules

Key symbols

Rule a (i):

The size of the reflected image is inversely proportional to the object's distance from the sphere center.

Increasing this distance reduces the size of the reflected image.

Rule a (ii):

Wherever the observer locates on a line between an object and the sphere, the reflected images of other stationary objects remain the same size.

Rule b:

The observer's eye or camera lens reflected on a spherical surface is always at the center of the sphere.

Rule c:

No matter what size the reflecting sphere is, the size of the reflected object is always in the same proportion to the sphere's size, given the same distances between the sphere and the object.

Alternatively, *The absolute magnitude of the reflected image is directly proportional to the sphere's diameter.*

Whatever size the reflecting sphere is, the reflected object always occupies the same proportion of the sphere. If your reflected body covers 50% of the sphere diameter, it is always 50% in any size sphere at the same distance from your body.

Note: These rules always stand while using small to medium size spheres. When using enormous spheres, where the sphere's radius is very similar to or larger than the distance between the sphere's surface and the observer, there will be variations in these rules. For example, using a 5-meter radius sphere with an observer very close to or beneath its surface.